The Meaning of Quantum Theory

The Meaning of Quantum Theory

A Guide for Students of Chemistry and Physics

Jim Baggott

OXFORD NEW YORK TOKYO

OXFORD UNIVERSITY PRESS

Oxford University Press, Walton Street, Oxford OX2 6DP

Oxford New York
Athens Auckland Bangkok Bombay
Calcutta Cape Town Dar es Salaam Delhi
Florence Hong Kong Istanbul Karachi
Kuala Lumpur Madras Madrid Melbourne
Mexico City Nairobi Paris Singapore
Taipei Tokyo Toronto
and associated companies in
Berlin Ibadan

Oxford is a trade mark of Oxford University Press

Published in the United States
by Oxford University Press Inc., New York

© Jim Baggott, 1992

First Published 1992
Reprinted 1994, 1995

A catalogue record for this book is available from the British Library

Library of Congress Cataloging in Publication Data
Baggott, J. E.
The meaning of quantum theory : a guide for students of chemistry
and physics / Jim Baggott.
Includes bibliographical references and index.
1. Quantum theory. 2. Quantum chemistry. I. Title.
QC174.12.B34 1992 503.1'2 — dc20 91-34937

ISBN 0 19 855575 X (Pbk)

Printed in Harrisonburg, Va. by
R.R.Donnelley & Sons Company

To Timothy

KERNER: Now we come to the exciting part. We will watch the bullets of light to see which way they go. This is not difficult, the apparatus is simple. So we look carefully and we see the bullets one at a time, and some hit the armour plate and bounce back, and some go through one slit, and some go through the other slit, and, of course, none go through both slits.

BLAIR: I knew that.

KERNER: You knew that. Now we come to my favourite bit. The wave pattern has disappeared! It has become particle pattern, just like with real machine-gun bullets.

BLAIR: Why?

KERNER: Because we looked. So, we do it again, exactly the same except now without looking to see which way the bullets go; and the wave pattern comes back. So we try again while looking, and we get particle pattern. Every time we don't look we get wave pattern. Every time we look to see how we get wave pattern, we get particle pattern. The act of observing determines the reality.

Tom Stoppard, *Hapgood*

Preface

Why have I written this book? Perhaps a more burning question for you is: Why should you read it?

I wrote this book because in August 1987 I made a discovery that shocked me. If, before this date, you had asked me at what stage in the process of emission and subsequent detection of a photon its state of polarization is established, I would have answered: At the moment of emission, of course! Imagine then that two photons emitted in rapid succession from an excited calcium atom are obliged, by the laws of atomic physics, to be emitted in opposite states of circular polarization: one left circularly polarized and one right circularly polarized. Surely, they set off from the atom towards their respective detectors already *in* those states of circular polarization. Yes?

Well, . . . no. I have since learned that this view — the assumption that the physical states of quantum particles like photons are 'real' *before* they are measured — is called (rather disparagingly, I sometimes think) naïve realism. Now in the 1920s and 1930s, some of the most famous figures in twentieth century physics were involved in a big debate about the meaning of the new quantum theory and its implications for physical reality. In August 1987 I knew a little bit about this debate. But I had assumed that it had the status of a philosophical debate, with little or no relevance to practical matters that could be settled in the laboratory. I had been trained as a scientist, and although I enjoyed reading about philosophy (like I enjoyed listening to music), I was too busy with more important matters to dig deeply into the subject.

In July and August 1987, I made a short study visit to the University of Wisconsin at Madison, where I bought a book (always dangerous) from the University bookstore. This was a book published in celebration of the centenary of the birth of Niels Bohr. In it were lots of articles about his contribution to physics and his great debate with Einstein on the meaning of quantum theory. One of these articles, written by N. David Mermin, gave me a tremendous shock. Mermin described the results of experiments that had been carried out as recently as 1982 to test something called Bell's theorem using two-photon 'cascade' emission

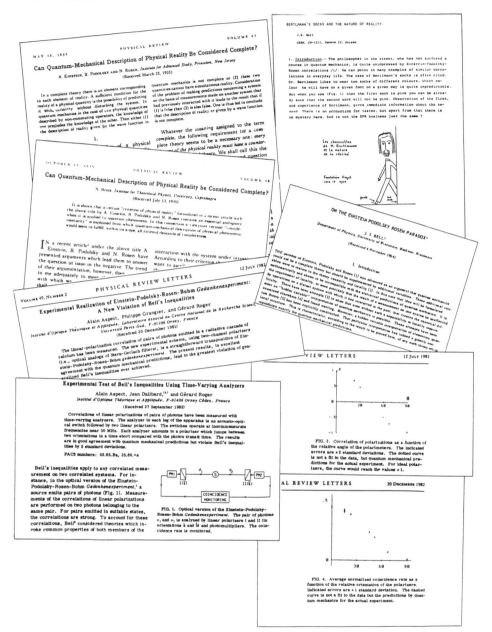

A pictorial history of the Einstein–Podolsky–Rosen paradox. Montage by P. J. Kennedy. Reproduced by permission of the publishers from French, A. P. and Kennedy, P. J. (eds) (1985). *Niels Bohr: a centenary volume*. Harvard University Press. MA. Copyright © 1985 by The International Commission on Physics Education.

from excited calcium atoms. Put simply, Bell's theorem says that my idea of naïve realism is in *conflict* with the predictions of quantum theory in a way that can be tested in the laboratory in special experiments on pairs of quantum particles. These experiments had been done: quantum theory had been proved right and naïve realism wrong! There in a montage was a pictorial history of the debate about reality and the experiments that had been done to test it (reproduced opposite).

This work struck me as desperately important to my understanding of physical reality, something that as a scientist I felt I ought to know about. This discovery also made me feel rather embarrassed. Here I was, proud of my scientific qualifications and with almost 10 years' experience in chemical physics research at various prestigious institutions around the world, and I had been going around with a conception of physical reality that was completely wrong! *Why hadn't somebody told me about this before?*

I could not rest until I had sorted all this out. How can it be that quantum particles are not 'real' until they are detected? Are alternative interpretations of quantum theory possible? If so, what are they like? I bought lots more books (some very expensive) and spent hours and hours trying to understand what was going on. There are many excellently written popular books on this subject that are easy to understand, but these left me dissatisfied. These books just told me that there is a problem, whereas I needed to know *why* there is a problem: to know what it is about the mathematics of quantum theory that leads to all these difficulties.

The trouble is that many of the most important works published on the interpretation of quantum theory are heavy going and I (with a mathematical background I will flatteringly describe as 'poor') made heavy weather of them. Nevertheless, I persevered and managed to arrive at something approaching comprehension. I decided to write it all out in a book, in such a way that undergraduate students of chemistry and physics should be able to comprehend the material without needing to spend hours poring over more advanced texts. And this, of course, answers the second question: this is why *you* should read this book.

Students of physical science are usually introduced to the subject matter of quantum theory at an early, sensitive period in their undergraduate studies. Sensitive, because their earlier instruction will not have taught them the whole truth about the nature of scientific activity and the way in which scientific progress is made. Sensitive also because they will not yet have been trained to question what they are told or what they read in textbooks.

Quantum theory is unnerving. Not only is the theory mathematically complex, it is also conceptually challenging. For the first time, students are taught about a theory which they have to accept and which they have to learn how to *apply*, but for which they cannot expect to be told its *meaning*. Many will not realize that their inability to understand the theory is due not to a failing on their part, but to the fact that quantum theory in its present form is inherently non-understandable. As Richard Feynman has said: '. . . I think I can safely say that *nobody* understands quantum mechanics'[†] (my italics).

Most undergraduate courses on quantum theory never touch on the theory's profound conceptual problems. This is because the theory brings us right back to some of the central questions of *philosophy* and, as we know, there is no room for philosophy in a modern science degree. I find this an absurd situation. It is my opinion, expressed in this book, that quantum theory *is* philosophy. Oh, we can dress it up in grand phrases littered with jargon—state vector, hermitian operator, Hilbert space, projection amplitude, and so on—we can make it all very mechanistic and mathematical and scientific, but this does not completely hide the truth. Beneath the formalism must be an interpretation, and the interpretation is pure philosophy.

This is the reason why quantum theory is such a difficult subject. Until they reach undergraduate level, students of chemistry and physics are brought up on classical science in which there appears to be no need for philosophy. They are consequently ill prepared to come to terms with quantum theory. And be warned: students are rarely told the whole truth about this theory. Instead they are fed the orthodox interpretation either by design or default. It is perhaps surprising that for a theory so fundamental to our understanding of much of chemistry and physics our teachers do not find it necessary to explain that it has many alternative interpretations. This is a great pity. Students have a right to know the truth, even if it is bizarre.

In this book I have posed five questions which I believe students might have expected to be provided with answers. These five questions are:

Why is quantum theory necessary?
How does it work?
What does it mean?
How can it be tested?
What are the alternatives?

I have tried to answer these questions, one in each of the book's five chapters. While I am not in a position to tell you what quantum theory means, I can tell you why its meaning is so elusive.

[†] Feynman, R. P. (1967). *The character of physical law*. M.I.T. Press, Cambridge, MA.

Thanks go to Mike Pilling, Ian Smith and Brian Elms for their constructive comments on early drafts of this book. Special thanks go to Peter Atkins, not only for the very helpful comments he made on the draft manuscript in his role as reviewer for Oxford University Press, but also for his excellent textbook *Molecular quantum mechanics* from which much of my knowledge of quantum theory and its applications has been derived. This book would not have been possible without the encouragement of my editor at OUP. I will remain eternally grateful for having an opportunity to get this lot off my chest.

JOHN S. BELL

It was with great sadness that I learned of the death of the physicist John S. Bell during my writing of this book. I had never met Bell, nor heard him lecture, but in my reading of his scientific papers I have developed a great admiration for him and his work. I have especially admired his attempts to dismantle the orthodox Copenhagen interpretation of quantum theory, written with such tremendous style and obvious enjoyment. Although in this book I have tried to present a balanced account — arguing one way and then another — I hope that I have done justice to Bell's superbly constructed criticisms. The debate over the meaning of quantum theory will certainly be poorer without him.

Reading
April 1991   J. E. B.

Contents

1
How quantum theory was discovered

1.1 AN ACT OF DESPERATION

A scientist in the late nineteenth century could be forgiven for thinking that the major elements of physics were built on unshakeable foundations and effectively established for all time. The inspirational work of Galileo Galilei and Isaac Newton in the seventeenth century had been shaped by a further 200 years of theoretical and, in particular, experimental science into a marvellous construction which we now call classical physics. This physics appeared to explain almost every aspect of the physical world: the dynamics of moving objects, thermodynamics, optics, electricity, magnetism, gravitation, etc. So closely did theory agree with and explain experimental observations of the everyday world that there could be no doubt about its basic correctness—its essential 'truth'. Admittedly, there were a few remaining problems but these seemed to be trivial compared with the fundamentals—a matter of dotting a few *i*s and crossing some *t*s.

And yet within 30 years these 'trivial' problems had turned the world of physics completely upside-down and, as we will see, subverted our cosy notions of physical reality. When extended to the microscopic world of atoms, the foundations of classical physics were shown to be not only shakeable, but built on sand. The emphasis changed. The physics of Newton was mechanistic, deterministic, logical and certain—there appeared to be little room for any doubt about what it all meant. In contrast, the new quantum physics was to be characterized by its indeterminism, illogicality and uncertainty; about 70 years after its discovery, its meaning remains far from clear.

It is sometimes difficult to understand how this could have happened. Why replace logic and certainty with illogic and uncertainty? There must have been very good reasons. If we are to accept what is implied by the new quantum physics, we must make the attempt to understand what these reasons were. Our guided tour of the meaning of quantum theory therefore begins with an examination of these reasons from a historical perspective. This is not intended to be a bland retelling of science history, but rather a good, hard look at how the early quantum theory developed and, most importantly, how that development was determined by the attitudes of the scientists involved: the early quantum theory's dramatis personnae. We begin with light.

Light at the turn of the century

By the end of the nineteenth century, overwhelming evidence had been accumulated in support of a wave theory of light. How else is it possible to explain light diffraction and interference? Almost a century earlier, Thomas Young demonstrated that the passage of light through two narrow, closely spaced holes or slits produces a pattern of bright and dark fringes (see Fig. 1.1). These are readily explained in terms of a wave theory in which the peaks and troughs of the waves from the two slits start out in phase. Where a peak of one wave is coincident with a peak of the other, the two waves add and reinforce (constructive interference), giving rise to a bright fringe. Where a peak of one wave is coincident with a trough of the other, the two waves cancel (destructive interference), giving a dark fringe. Despite the logic of this explanation, it was rejected by the physics community at the time Young proposed it. Newton's corpuscular theory of light had dominated physics since the seventeenth century and had become something of a dogma; arguments against it were not readily accepted.

Perhaps the most conclusive evidence in favour of a wave theory of light came in the 1860s from James Clerk Maxwell's work on electricity and magnetism. Following the marvellous experimental work of Michael Faraday, Maxwell combined electricity and magnetism in a single theory. He proposed the existence of electromagnetic fields whose

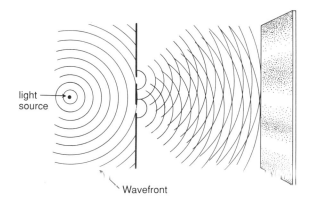

Fig. 1.1 Light interference in a double-slit apparatus.

properties are described by his theory. He made no assumptions about how these fields move through space. Nevertheless, the mathematical form of Maxwell's equations — equations that connect the space and time dependences of the electric and magnetic components of the fields — point unambiguously to a wave-like motion.

The equations also indicate that the speed of the waves should be a constant, related to the permittivity and permeability of free space. When Maxwell calculated what this constant speed was predicted to be, he found it[†]

. . . so nearly that of light, that it seems we have strong reason to conclude that light itself (including radiant heat, and other radiations if any) is an electromagnetic disturbance in the form of waves propagated through the electromagnetic field according to electromagnetic laws.

Furthermore, for one-dimensional plane waves, Maxwell's equations do not allow the component of the field in the direction of propagation to vary. In other words, plane electromagnetic waves (and hence plane polarized light waves) are transverse waves; they oscillate at right angles to the direction in which they are moving, as Young had proposed about 40 years earlier. An example of such a plane wave is shown in Fig. 1.2.

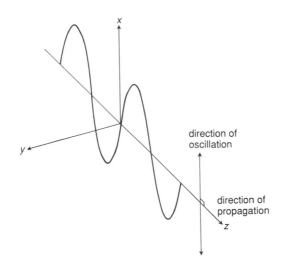

Fig. 1.2 A plane wave — the wave oscillates at right angles to the direction of propagation.

[†] Quotation from Hecht, Eugene and Zajac, Alfred (1974). *Optics*. Addison-Wesley, Reading, MA.

A few difficulties remained, however. For example, all wave motion requires a medium to support it, and the so-called luminiferous ether — supposedly a very tenuous form of matter — was the favoured medium for light waves. But if the existence of the ether was accepted, certain physical consequences had to follow. The earth's motion through a motionless ether should give rise to a drag effect and hence there should be measureable differences in the speed of light depending on the direction it is travelling relative to the earth. This idea was put to its most stringent test by Albert Michelson and Edward Morley in 1887. They found no evidence for a drag effect and hence no evidence for relative motion between the earth and the ether. This is one of the most important 'negative' experiments ever performed, and led to the award of the 1907 Nobel prize in physics to Michelson.

But there was another, seemingly innocuous, phenomenon involving light that was causing physicists some problems at the end of the nineteenth century. This was the problem of black-body radiation, and solving it led to the development of quantum theory.

Black-body radiation and the ultraviolet catastrophe

When we heat an object to very high temperatures, it absorbs energy and emits light. We use phrases such as 'red hot' or 'white hot' to describe this effect. Different objects tend to emit more light in some frequency regions than in others. A black body is one of those model objects invented by theoretical physicists which are good approximations of real objects but which are theoretically easier to describe. A black body absorbs and emits radiation perfectly, i.e. it does not favour any particular range of radiation frequencies over another. Thus, the intensity of radiation emitted is directly related to the amount of energy in the body when it is in thermal equilibrium with its surroundings.

The theory of black-body radiation has a fascinating history, not only because it encompasses the discovery of quantum theory but also because its development is so typical of the frequently tortuous paths scientists follow to sometimes new and unexpected destinations. Theoretical physicists realized that they could develop a theory of black-body radiation by studying the properties of radiation trapped inside a cavity. This is simply a box with perfectly insulating walls which can be heated and which is punctured with a small pinhole through which radiation can enter and leave. The radiation observed through the pinhole when the cavity is in thermal equilibrium is then equivalent to that of a perfect black body.

The theoreticians devised models for black-body radiation based on vibrations or oscillations of the electromagnetic field trapped inside a radiation cavity. These vibrations were assumed to be caused by the interaction between the electromagnetic field and a set of oscillators of a largely unspecified nature. We would now identify these oscillators as the constituent atoms of the material from which the cavity is made. Energy is released from excited atoms in the form of light (ultraviolet, visible and infrared, depending on the temperature), and the cavity eventually achieves an equilibrium — a dynamic balance between energy absorption and emission. However, remember that in the latter half of the nineteenth century there was still much uncertainty about the reality of atoms and molecules and J. J. Thomson's experiments confirming the existence of electrons were not performed until 1897.

It was imagined that as the external temperature of a cavity is increased, so the distribution of the frequencies of the oscillators shifts to higher ranges. This in turn causes vibrations in the electromagnetic field of higher and higher frequency, with a certain oscillator frequency giving rise to the same frequency of vibration in the field. These vibrations were visualized as standing waves: waves which 'fit' exactly in the space between the walls of the cavity and which were reinforced by constructive interference.

Early experimental studies established that the emissivity of a black body — a measure of its emissive power — is a function of frequency and temperature. In our discussion here we will make use of a property called the spectral (or radiation) density, $\rho(\nu, T)$, which is the density of radiation energy per unit volume per unit frequency interval $d\nu$ at a temperature T. In 1860, Gustav Kirchhoff challenged the scientific community to discover the functional form of the dependence of $\rho(\nu, T)$ on frequency and temperature. Towards the end of the nineteenth century, breakthroughs in the experimental study of, in particular, infrared radiation emitted from a radiation cavity allowed the models developed by the theoreticians to be tested stringently.

Models based on the general principles described above had been proposed which allowed $\rho(\nu, T)$ to be calculated for given values of ν and T. These expressions were moderately successful, but tended to fail at the extremes of frequency. For example, in 1896, Wilhelm Wien used a simple model (and made some unjustified assumptions) to derive the expression

$$\rho(\nu, T) = \alpha \nu^3 e^{-\beta \nu/T} \qquad (1.1)$$

where α and β are constants. This seemed to be quite acceptable, and was supported by the experiments of Freidrich Paschen in 1897. However, new experimental results obtained by Otto Lummer and Ernst Pringsheim reported in 1900 showed that Wien's formula failed in the low frequency infrared region.

In June 1900 Lord Rayleigh published details of a theoretical model based on the 'modes of etherial vibration' in a radiation cavity. Each mode possessed a specific frequency, and could take up and give out energy continuously. Rayleigh assumed a classical equipartition (or distribution) of energy over these modes. Such a distribution requires that, at equilibrium, each mode of vibration should possess an energy of kT, where k is Boltzmann's constant. Rayleigh duly arrived at the result $\rho(\nu, T) \propto \nu^2 T$. In May 1905, he obtained an expression for the constant of proportionality, but made an error in his calculation which was put right by James Jeans the following July and the result, known as the Rayleigh–Jeans law, can be written:

$$\rho(\nu, T) = \frac{8\pi\nu^2}{c^3} kT. \tag{1.2}$$

This expression was quite successful at low frequencies, where Wien's law failed, but it is obviously also ridiculous. It implies that the spectral density should increase in proportion to ν^2 without limit and so the total energy emitted, which is given by the integral of $\rho(\nu, T)$ with respect to ν, should be infinite. Because the theory predicts an accumulation of energy at high radiation frequencies, in 1911 the Austrian physicist Paul Ehrenfest called this problem the 'Rayleigh–Jeans catastrophe in the ultraviolet', now commonly known as the ultraviolet catastrophe.

Between 1900 and 1905 the German physicist Max Planck had arrived at a very successful radiation formula, described below. However, many physicists had regarded Planck's formula as providing merely an empirical 'fit' to the experimental data, and to be without theoretical justification. The ultraviolet catastrophe caused them to look more closely at Planck's result.

Planck's radiation formula

It has been suggested that Planck discovered his radiation formula on the evening of 7 October 1900. He had been paid a visit at his home in Berlin by the physicist Heinrich Reubens, who told him of some new experimental results he had obtained with his colleague Ferdinand Kurlbaum. They had studied black-body radiation even further into the infrared than Lummer and Pringsheim and had found that $\rho(\nu, T)$ becomes proportional to T at low frequencies (as required by the Rayleigh–Jeans law, although at the time Planck was not aware of Rayleigh's June 1900 paper). By combining this information with Wien's earlier expression, Planck deduced an expression which fitted all the available experimental data. He obtained

$$\rho\left(\nu, T\right) = \frac{8\pi\nu^2}{c^3} \frac{h\nu}{e^{h\nu/kT} - 1}. \tag{1.3}$$

This expression reproduces Wien's formula for high frequencies ($h\nu/kT \gg 1$), and at low frequencies ($h\nu/kT \ll 1$) it reproduces the Rayleigh–Jeans law. The constant h (Planck's constant) is related to Wien's α, and Wien's $\beta = h/k$.

Planck proposed his radiation formula at a meeting of the German Physical Society on 19 October 1900. The next day, Reubens compared his experimental results with Planck's formula and found the agreement to be 'completely satisfactory'.

Having obtained his formula, Planck was concerned to discover its physical basis. After all, he had arrived at his result somewhat empirically and was keen to derive the formula using more rigorous methods. He chose to approach the problem through thermodynamics. Using basic thermodynamics, he derived an expression for the entropy S of an oscillator in terms of its internal energy U and its frequency of oscillation:

$$S = k \left[\left(1 + \frac{U}{h\nu}\right) \ln \left(1 + \frac{U}{h\nu}\right) - \frac{U}{h\nu} \ln \frac{U}{h\nu} \right]. \tag{1.4}$$

The interested reader may find an explanation of Planck's derivation in Appendix A.

Equation (1.4) is an expression for the entropy of an oscillator that is consistent with Planck's radiation formula and therefore consistent with experiment. Thus, the physical basis of the radiation formula would be established if a second, theoretical, expression for S could be derived more directly from the intrinsic properties of the oscillators themselves.

At the time that Planck was struggling to find an alternative derivation, the Austrian physicist Ludwig Boltzmann had long advocated a new approach to the calculation of thermodynamic quantities using statistics. Planck did not like Boltzmann's statistical approach at all, but he was forced to use it. As he later explained in a letter to Robert Williams Wood:[†]

. . . what I did can be described as simply an act of desperation . . . A theoretical interpretation [of the radiation formula] . . . had to be found at any cost, no matter how high.

[†] Planck, Max, letter to Wood, Robert Williams, 7 October 1931.

Boltzmann's statistical approach

Ludwig Boltzmann was an adherent of the strictly mechanical approach to interpreting and understanding physical phenomena, a viewpoint which we will discuss in a little more detail in Chapter 3. In 1877, he developed his own entirely mechanical interpretation of the second law of thermodynamics. Entropy, he argued, is simply a measure of the probability of finding a mechanical system composed of discrete atoms or molecules in a particular state. The second law is therefore a general statement that a system with a low probability (low entropy) will evolve in time into a state of higher probability (higher entropy). The equilibrium state of a system is the one of highest probability, i.e. it is the most likely state.

In applying Boltzmann's ideas to the theory of black-body radiation, Planck had to assume that the total energy could be split up into a collection of indistinguishable but independent elements (or 'packets'), each with an energy ε, which were then statistically distributed over a large number of distinguishable oscillators. Planck may have had more than half an eye on the result that he was aiming for, because in making the energy elements *indistinguishable* he was following a very different path from Boltzmann. In his excellent biography of Albert Einstein, *Subtle is the Lord*, Abraham Pais wrote that[†] 'From the point of view of physics in 1900 the logic of Planck's electromagnetic and thermodynamic steps was impeccable, but his statistical step was wild.'

In 1911, Paul Ehrenfest demonstrated that Planck's statistical approach implied the existence of 'particles' of energy unlike any that had ever been invoked before. As we will see, Ehrenfest was right to be suspicious – Planck's particles of energy were not like just any other particles. Apart from this aspect of Planck's derivation, the remainder relied on standard methods of statistical thermodynamics. For those readers interested in following Planck's reasoning, his derivation is given in Appendix A. We proceed here directly to his result:

$$S = k \left[\left(1 + \frac{U}{\varepsilon} \right) \ln \left(1 + \frac{U}{\varepsilon} \right) - \frac{U}{\varepsilon} \ln \frac{U}{\varepsilon} \right]. \qquad (1.5)$$

By comparing eqns (1.4) and (1.5), Planck was led to the unavoidable conclusion that the finite energy elements ε had the form

$$\varepsilon = h\nu. \qquad (1.6)$$

The world of physics would never be the same again.

[†] Pais, Abraham (1982). *Subtle is the Lord: the science and the life of Albert Einstein*. Oxford University Press.

Planck was a very reluctant scientific revolutionary. Although his radiation formula could be derived from 'first principles' using Boltzmann's statistical approach, he did not really like the idea that energy could be taken up or given out by the oscillators only in discrete elements (which he later called quanta). Newtonian physics said that energy was continuously variable, and yet the statistical approach seemed to suggest that energy must be 'quantized'.

Although Planck eventually became an adherent of Boltzmann's statistical theory he believed for some time that his 'solution' to the problem of black-body radiation held no deeper meaning, other than that of giving the correct result.

Quanta

Planck was not the only one to have mixed feelings about his interpretation of the radiation formula; most of the physics community was sceptical. While most physicists acknowledged the fact that Planck's radiation formula gave the correct result, some found it hard to believe that energy could be quantized. A few physicists initially believed that Planck's interpretation was so monstrous that the radiation formula itself (and hence also the experimental results) must be wrong.

However, the seeds of the quantum revolution had been sown. Planck's work was studied carefully by a 'technical expert, third class' in the Swiss Patent Office in Bern. His name was Albert Einstein. In 1905, Einstein expressed reservations about Planck's derivation, pointing out that Planck had been inconsistent in first assuming energy to be continuously variable and then assuming exactly the opposite when comparing eqn (1.4) with eqn (1.5) by setting $\varepsilon = h\nu$.

But, unlike most other physicists, Einstein was prepared to accept the reality of quanta. His genius was to accept the 'impossible' and use it to explain other puzzling phenomena, making predictions that could be tested experimentally. In a paper published in 1905, Einstein introduced his light-quantum hypothesis:[†]

Monochromatic radiation . . . behaves . . . as if it consists of mutually independent energy quanta of magnitude [$h\nu$].

He went even further, suggesting that[†]

If . . . monochromatic radiation . . . behaves as a discrete medium consisting of energy quanta of magnitude [$h\nu$], then this suggests an inquiry as to whether

---

[†] Einstein, A. (1905). *Annalen der Physik*. **17**, 132.

the laws of the generation and conversion of light are also constituted as if light were to consist of energy quanta of this kind.

In other words, Einstein was prepared not only to embrace the idea of light quanta (which were called photons by G. N. Lewis in 1926), but also to look at its implications.

Nearly 90 years later, it is difficult to imagine just how revolutionary Einstein's ideas were. They were not readily accepted by most physicists at the time, but the ultraviolet catastrophe, and other problems which we will mention briefly in the next section, eventually convinced them of the need for the quantum hypothesis.

1.2 GATHERING THE EVIDENCE

Science is a democratic activity. It is rare for a new theory to be adopted by the scientific community overnight. Rather, scientists need a good deal of persuading before they will invest belief in a new theory; especially if it provides an interpretation that runs counter to their intuition, built up after a long acquaintance with the old way of looking at things. This process of persuasion must be backed up by hard experimental evidence, preferably from new experiments designed to test the predictions of the new theory. Only when a large cross-section of the scientific community believes in the new theory is it accepted as 'true'.

So it was with quantum theory. Although Einstein proposed his light-quantum hypothesis in 1905, it took about 20 years of hard work by both theoreticians and experimentalists before it was widely accepted. The resistance of many physicists to these new ideas is understandable: quantum theory was like nothing they had ever seen before.

The photoelectric effect

The theory scored some notable early successes, largely through Einstein's inspired efforts. In 1905, Einstein used his light-quantum hypothesis to explain the photoelectric effect. This was another effect that had been puzzling physicists for some time. It was known that shining light on metal surfaces could lead to the ejection of electrons from these surfaces. However, contrary to the expectations of classical physics, the kinetic energies of the emitted electrons show no dependence on the intensity of the radiation, but instead vary with the radiation frequency. This is strange because the energy contained in a classical wave depends on its amplitude (and hence its intensity), not its frequency.

Einstein solved this problem by suggesting that a light-quantum incident on the surface transfers all of its energy to a single electron. That electron is ejected with a kinetic energy equal to the energy of the light-quantum less an amount expended by escaping to the surface and which is therefore characteristic of the metal (a property now known as the work function).

According to Planck, the energy of the light-quantum is given by $\varepsilon = h\nu$, and so the kinetic energy of the ejected electron is expected to increase with increasing frequency. Increasing the intensity of the radiation increases the number of light quanta incident on the surface, increasing the number, but not the kinetic energies, of ejected electrons. Einstein's theory was very simple, and yet it made a number of important, testable predictions. These were confirmed in a series of experiments performed about 10 years later. Einstein's work on the photoelectric effect won him the 1921 Nobel prize for physics.

Bohr's theory of the atom

In June 1912, the Danish physicist Niels Bohr wrote to his brother Harald, telling him that he believed he had 'found out a little' about atoms that might represent 'perhaps a little bit of reality'. Bohr was working in Manchester with Ernest Rutherford on the problems of atomic structure. Rutherford had earlier demonstrated experimentally that atoms consist of a massive, positively charged nucleus surrounded by much less massive electrons. Bohr became convinced that the origin of the chemical properties of an element was to be found in the properties of the electron system surrounding its nucleus.

At the end of 1912, Bohr came across J.W. Nicholson's quantum model of the atom and, like Nicholson, he became concerned to find an explanation for why atoms, particularly the hydrogen atom, absorbed and emitted radiation only at certain discrete, well defined frequencies. In 1885, Balmer had measured one series of hydrogen emission lines and found them to follow a relationship which became known as the Balmer formula:

$$\nu_n = R\left\{ \frac{1}{2^2} - \frac{1}{n^2} \right\} \quad n = 3, 4, 5, \ldots \qquad (1.7)$$

where ν_n is the frequency of the emitted radiation and R became known as the Rydberg constant. It was the involvement of the integer numbers n that gave Bohr a clue to the explanation of Balmer's formula.

Bohr developed a theory of the atom in which the electrons move around the nucleus in fixed, stable orbits much like the planets orbit the sun. In terms of classical physics, such a model is impossible. A charged particle moving in an electrostatic field radiates energy. An orbiting electron would therefore be expected to lose energy continuously, eventually spiralling into the nucleus. Nevertheless, Bohr postulated that, despite the apparent inconsistency, experiment revealed the existence of stable electron orbits.

By fusing this 'impossible' classical mechanical picture with Planck's quantum theory, Bohr was able to argue that only certain orbits are 'allowed', and an electron moving from an outer, higher-energy orbit to a lower-energy orbit causes the release of energy as emitted radiation. Because the orbits are fixed, so are the energy gaps between them and hence atomic emission can be observed only at those radiation frequencies corresponding to the energy gaps. However, now Bohr faced an even bigger problem than he had set out to solve. Even in Planck's radiation theory, the frequency of radiation released from an atomic oscillator is dependent on the mechanical frequency of the electron producing it. Bohr had to propose that this is no longer acceptable: the radiation frequency must differ from the frequency of the oscillator.

Bohr published his theory of the atom in 1913 in a series of papers. Try to imagine the state of physics at the time, with physicists still uncertain about Planck's interpretation of his radiation law, with few caring overmuch for Einstein's light-quantum hypothesis and a great deal of confusion around, and you will get some idea of Bohr's breathtaking vision.

In the first of three papers setting out his new theory, Bohr adopted a model for the hydrogen atom based on an electron forced to move in a stable elliptical orbit around a singly positively charged nucleus. From this model, he obtained an expression for the mechanical angular frequency of the electron moving in such an orbit. Bohr then used this result to determine the amount of energy emitted by the atom when the electron is brought into one of the stable orbits from a great distance away from the nucleus. He obtained (in modern notation):

$$E_n = -\frac{m_e e^4}{8h^2 \varepsilon_0^2 n^2}. \tag{1.8}$$

E_n is the energy associated with the formation of the stable orbit characterized by the integer number n (later to become known as the quantum number); m_e is the mass of the electron, e is the electron charge and ε_0 is the vacuum permittivity. E_n is negative since it represents a state of lower energy compared with the completely separated stationary electron and nucleus defined as the arbitrary energy zero.

The energy emitted by the atom as an electron falls from a high energy orbit (characterized by the quantum number n_2) to a lower energy orbit (quantum number n_1) is therefore given by:

$$E_{n_2} - E_{n_1} = \frac{m_e e^4}{8h^2 \varepsilon_0^2} \left\{ \frac{1}{n_1^2} - \frac{1}{n_2^2} \right\}. \tag{1.9}$$

Bohr then supposed that this energy is released as radiation with frequency ν, i.e.

$$E_{n_2} - E_{n_1} = h\nu, \qquad (1.10)$$

and hence

$$\nu = \frac{m_e e^4}{8h^3 \varepsilon_0^2} \left\{ \frac{1}{n_1^2} - \frac{1}{n_2^2} \right\}. \qquad (1.11)$$

Thus, Balmer's formula is just a special case of a more general expression with $n_1 = 2$ and $n_2 = 3$, 4, 5, etc and the Rydberg constant is a collection of fundamental physical constants. Bohr noted that $n_1 = 3$ gives the Paschen series, and that $n_1 = 1$ and $n_1 = 4$ and 5 predicted further series in the ultraviolet and infrared that, at that time, had not been observed.

A further series of emission lines known as the Pickering series was thought by experimental spectroscopists also to belong to the hydrogen atom. However, at the time, the Pickering series was characterized by half-integer quantum numbers which are not possible in Bohr's theory. Instead, Bohr proposed that the formula be rewritten in terms of integer numbers, suggesting that the Pickering series belongs not to hydrogen atoms but to ionized helium atoms. An awkward mismatch between calculated and observed emission frequencies was later resolved by Bohr when he realized that he had neglected the effect of the motion of the heavy helium nucleus on the stable electron orbits of ionized helium. This correction gave a Rydberg constant for ionized helium some 4.00163 times greater than that for hydrogen (not 4 times greater, as Bohr had originally proposed). The experimentalists found this ratio to be 4.0016. When he heard about this result, Einstein described Bohr's theory as 'an enormous achievement'.

In arriving at eqn (1.8), Bohr had had to assume that the kinetic energy of an electron moving in an elliptical orbit around the nucleus is equal to half the potential energy. Bohr tried to justify this using arguments based on the properties of an atomic oscillator of the type Planck had used in his derivation of the radiation law. However, he later abandoned this argument in favour of one originally developed by Nicholson: this relationship follows from the fact that the orbital angular momentum of an electron moving around the nucleus in a *circular* orbit is a fixed quantity with a value of $h/2\pi$.

Bohr's idea of stable electron orbits had a further consequence. Transitions between the orbits had to occur in instantaneous 'jumps', because if the electron gradually moves from one orbit to another, it would again be expected to radiate energy continuously during the process. This is certainly not what is observed when an atom absorbs light. Thus,

transitions between inherently non-classical stable orbits must themselves involve non-classical discontinuous 'quantum jumps'. Bohr wrote
that[†] 'the dynamical equilibrium of the systems in the [stable orbits]
is governed by the ordinary laws of mechanics, while these laws do
not hold for the passing of the systems between the different [stable
orbits].' Perhaps surprisingly, at this stage Bohr did not believe in light
quanta.

Spontaneous emission

It is worthwhile noting that, from almost the very beginning, Einstein
viewed the quantum interpretation as provisional, to be eventually
replaced by a new, more complete theory that would explain quantum
phenomena somewhat more rigorously. Einstein's attitude towards the
new quantum physics, and his celebrated debate with Niels Bohr on the
meaning of the theory, are discussed in detail in Chapter 3.

In 1916 and 1917, Einstein published his work on the spontaneous and
stimulated emission of radiation by molecules (and incidentally laid the
foundations of the theory of the laser). Einstein noted that the timing of
a spontaneous transition, and the direction of the consequently emitted
light-quantum, could not be predicted using quantum theory. In this
sense, spontaneous emission is like radioactive decay. The theory allows
the calculation of the *probability* that a spontaneous transition will take
place, but leaves the exact details entirely to *chance*. (We will look into
this in more detail in Chapter 2.) Einstein was not at all comfortable with
this idea. Three years later he wrote to Max Born on the subject of the
absorption and emission of light, noting that he 'would be very unhappy
to renounce complete causality'.[‡] After pioneering quantum theory
through one of its most testing early periods, Einstein was beginning to
have doubts about the theory's implications. These doubts were to turn
Einstein into one of the theory's most determined critics.

We are today so used to the notion of a spontaneous transition that
it is, perhaps, difficult to see what Einstein got so upset about. Let me
propose the following (very imperfect) analogy. Suppose I lift an apple
three metres off the ground and let go. This represents an unstable situation with respect to the state of the apple lying on the ground, and so
I expect the force of gravity to act immediately on the apple, *causing*
it to fall. Now imagine that the apple behaves like an excited electron

[†] Bohr, N. (1913). *Philosophical Magazine*. Reproduced in French, A.P. and Kennedy,
P.J. (eds.) (1985). *Niels Bohr: a centenary volume*. Harvard University Press, Cambridge,
MA.
[‡] Einstein, Albert, letter to Born, Max. 27 January 1920.

in an atom. Instead of falling back as soon as the 'exciting' force is removed, the apple hovers above the ground, falling at some unpredictable moment that I can calculate only in terms of a probability. Thus, there may be a high probability that the apple will fall within a very short time, but there may also be a distinct, small probability that the apple will just hover above the ground for several days!

We must be a little careful in our discussion of causality. An excited electron *will* fall to a more stable state; it is *caused* to do so by the quantum mechanics of the electromagnetic field. However, the exact moment of the transition appears to be left to chance. In quantum theory, the direct link between cause and effect appears to be severed.

The Compton effect

By 1909, Einstein imagined radiation to be composed of 'pointlike quanta with energy $h\nu$', a clear reference to a particle description. However, one unambiguous way of demonstrating that something has a particle nature is to try to hit something else with it. The first 'something else' was an electron. In 1923, Arthur Compton and Peter Debye both used simple conservation of momentum arguments to show that 'bouncing' light quanta off electrons should change the frequencies of the quanta by readily calculable amounts. Compton compared his prediction with experiment, and concluded that a light-quantum has a directed momentum, like a small projectile. The theory of light had come full circle; more than 200 years after Newton, light was once again thought to consist of particles.

But this was not a return to Newton's corpuscular theory. Experiments demonstrating the unambiguously wave-like properties of light, and their intepretation by Young in terms of waves, were not invalidated by the Compton effect. Likewise, the electromagnetic theory created by the work of Faraday and Maxwell was not torn down. Instead, physicists had to confront the difficult task of somehow fusing together the wave-like and particle-like aspects of light in a single, coherent theory. That theory had to be based on the essential dual wave–particle properties of light quanta.

1.3 WAVE–PARTICLE DUALITY

Einstein's special theory of relativity

Einstein introduced his special theory of relativity in 1905. He had struggled, and failed, to find a way of accomodating two general observations—the absence of an ether and an apparently universal speed of light, independent of the relative motion of the source—in any kind of Newtonian interpretation of space and time. Instead, he decided to accept these observations at face value and developed a new theory from the bare minimum of assumptions (or postulates) in which they would automatically result. He found that he needed only two.

He postulated that the laws of physics should be completely objective, i.e. they should be identical for all observers. In particular, they should not depend in any way on how an observer is moving relative to an observed object. In practical terms, this means that the laws of physics should appear to be identical in any so-called inertial frame of reference and so all such frames of reference are equivalent. An observer stationary in one frame of reference should be able to draw the same conclusions from some set of physical measurements as another observer moving relative to the first (or stationary in his own moving frame of reference). Einstein also postulated that the speed of light should be regarded as a universal constant, representing an ultimate speed which cannot be exceeded. (The fact that this speed happens to be that of light is irrelevant—light happens to travel at the ultimate speed.)

Unfortunately, we have no time in this book to examine the bizarre consequences of the special theory of relativity. Out went any idea of an absolute frame of reference (and hence the idea of a stationary ether), together with absolute space, time and simultaneity. In came all sorts of strange effects predicted for moving objects and clocks within a new four-dimensional space–time, all later confirmed by experiment. However, it is worthwhile noting that although the predictions of special relativity are rather strange, the theory is really one of classical physics.

For our present purposes, all we need at this stage is to note that for the kinetic energy of a freely moving particle, the demands of special relativity are met by the equation

$$E^2 = p^2 c^2 + m_0^2 c^4 \tag{1.12}$$

where p is the linear momentum of the particle, m_0 is its rest mass, and c is the speed of light. A particle moving with velocity v has an inertial mass m given by the equation

$$m = \frac{m_0}{(1 - v^2/c^2)^{1/2}}. \tag{1.13}$$

Thus, as v approaches c, the inertial mass m (a measure of the particle's resistance to acceleration) tends to infinity and it becomes increasingly difficult to accelerate the particle further. Equation (1.13) demonstrates the role of c as an ultimate speed.

A photon (with energy ε) moves at the speed of light and is thought to have zero rest mass. Thus, eqn (1.12) reduces to:

$$\varepsilon = pc \qquad (1.14)$$

where we have taken the positive root (more generally, $\varepsilon = |p|c$).

De Broglie's hypothesis

In 1923, the French physicist Louis de Broglie combined the results of Einstein's special theory of relativity and Planck's quantum theory to produce a new, 'tentative' theory of light quanta. Although he supposed that a light-quantum possesses a small rest mass, we can obtain his result simply by combining eqns (1.6) and (1.14):

$$\varepsilon = h\nu = pc . \qquad (1.15)$$

Since $\nu = c/\lambda$, where λ is the wavelength of the light-quantum, eqn (1.15) can be rearranged to give an expression for λ in terms of p:

$$\lambda = \frac{h}{p} . \qquad (1.16)$$

This is the de Broglie relation.

De Broglie went further. He suggested that this relation should hold for any moving particle with linear momentum p, and that moving particles should therefore exhibit corresponding wave-like properties characterized by a wavelength. In particular, he suggested that a beam of electrons could be diffracted.

That this wave nature of particles is not apparent in macroscopic objects, like high velocity bullets, is due to the very small size of Planck's constant h. If Planck's constant were very much larger, the macroscopic world would be an even more peculiar place that it is (the physicist George Gamow has speculated on what it might be like to play quantum billiards). However, because Planck's constant is so small, the dual wave–particle nature of matter is apparent only in the microscopic world of the fundamental particles. Of course, if Planck's constant were zero, there would be no duality, the world would be entirely 'classical' and I wouldn't be writing this book!

De Broglie collected his published papers together and presented them to his research supervisor, Paul Langevin, as a PhD thesis. Langevin sent a copy to Einstein and asked him for his views on it. Einstein wrote back saying that he found de Broglie's approach 'quite interesting'. Consequently, Langevin was happy to accept de Broglie's thesis, which was eventually published in its entirety in the journal *Annales de Physique* in

1925. This work was to have an important influence on the Austrian physicist Erwin Schrödinger.

Einstein and Bohr in conflict

Before we go on to find out just how de Broglie's ideas of wave–particle duality led to Schrödinger's wave mechanics, let us take a brief look at one of the very earliest episodes in what was to become a great debate between Einstein and Bohr on the meaning of quantum theory. Bohr and Einstein first met in 1920, and developed a strong friendship. However, in 1924 Bohr, in collaboration with Hendrik Kramers and John Slater, published a paper that contained proposals that alarmed Einstein, to the extent that Einstein regarded himself to be in *conflict* with Bohr. It was a conflict that was to have a profound impact on the further development of quantum theory and its interpretation.

Bohr did not like the idea of the light-quantum, and this dislike led him to develop a new approach to light absorption and emission by atoms. Bohr, Kramers and Slater (BKS) proposed that the 'sudden leaps' (quantum jumps) associated with light absorption and emission meant that the ideas of energy and momentum conservation had to be abandoned. Einstein had thought of taking such a step himself about 10 years earlier, but had finally decided against it. What alarmed Einstein most of all, however, was a further proposal that the idea of strict causality should also be abandoned. As we mentioned earlier, Einstein had already felt very uneasy about the element of chance implied in spontaneous emission — that a light-quantum could be ejected from an atom or molecule at some unpredictable moment determined by no apparent cause.

Although BKS suggested that there was no such thing as a truly spontaneous transition, their solution was to embrace the idea that probabilistic laws, involving so-called 'virtual' fields working in a non-causal manner, are responsible for inducing the transition. The BKS proposals immediately came under fire from all sides. They led to further experimental work on the Compton effect which clearly demonstrated that energy and momentum are indeed conserved. When the accumulated evidence against the BKS theory was overwhelming, Bohr promised to give their 'revolutionary' efforts a decent burial, and managed to overcome his resistance to the light-quantum. However, Bohr remained convinced that the quantum theory still demanded a new, revolutionary interpretation. The stage was set for a debate on the meaning of quantum theory between Bohr and Einstein that was to be one of the most remarkable debates in the history of science.

Postscript: electron diffraction and interference

De Broglie suggested in 1923 that the wave-like nature of electrons could be demonstrated by the diffraction of an electron beam through a narrow aperture. Earlier, in 1912, the demonstration by Max von Laue of the diffraction of X-rays by crystals was quickly developed into a powerful analytical tool for determining crystal and molecular structures.

In 1925, Clinton Davisson and Lester Germer (accidentally!) obtained an electron diffraction pattern from large crystals of nickel. In the same year, G.P. Thomson and A. Reid demonstrated electron diffraction by passing beams of electrons through thin gold foils. Davisson and Thomson shared the 1937 Nobel prize for physics for their work on the wave properties of electrons. In a nice twist of history, G.P. Thomson won the Nobel prize for showing that the electron is a wave whereas, 31 years earlier, his father J.J. Thomson had been awarded the Nobel prize for showing that the electron is a particle! Today, electron diffraction is used routinely to determine the structures of molecules in the gas phase.

The wave-like nature of electrons should also give rise to interference effects analogous to those described for light by Thomas Young. Double-slit interference of a beam of electrons has long been discussed by physicists, but was demonstrated in the laboratory for the first time only in 1989. The interference patterns obtained are shown in Fig. 1.3. In this sequence of photographs, each white spot registers the arrival of an electron that has passed through a double-slit apparatus. With a few electrons, it is impossible to pick out any pattern in the spots—they seem to appear randomly. But as their number is increased a clear interference pattern, consisting of 'bright' and 'dark' fringes, becomes discernible.

The appearance of distinct spots suggests that each individual electron has a particle-like property (each spot says an electron struck here), and yet the interference pattern is obviously wave like. In anticipation of some fun to come in Chapters 2 and 3, you might like to imagine what happens to an individual electron as it passes through the double-slit apparatus.

1.4 WAVE MECHANICS

On 23 November 1925, Erwin Schrödinger gave a presentation on de Broglie's thesis work at a seminar organized by physicists from the University of Zürich and the Eidgenössische Technische Hochschule. In

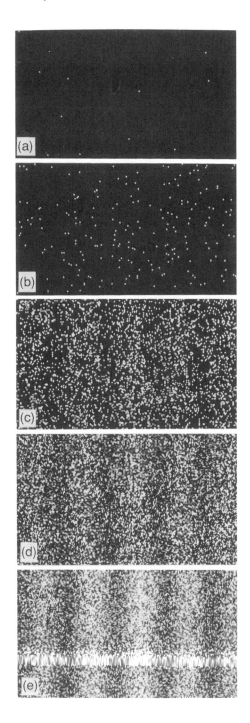

the discussion that followed, Peter Debye commented that he thought this approach to wave–particle duality to be somewhat 'childish'. After all, said Debye, 'to deal properly with waves one had to have a wave equation . . .'†

A few days before Christmas, Schrödinger left Zürich for a vacation in the Swiss Alps, leaving his wife behind but taking an old girlfriend (Schrödinger was noted for his womanizing) and his notes on de Broglie's hypothesis. We do not know who the girlfriend was or what influence she might have had on him, but when he returned on 9 January 1926, he had discovered wave mechanics.

How to 'derive' the Schrödinger equation

It is in fact impossible to derive (with any rigour) the quantum mechanical Schrödinger equation from classical physics. In many textbooks on quantum theory, the equation is simply given and then justified through its successful application to systems of interest to chemists and physicists. However, the equation had to come from somewhere, and it is indeed possible to 'derive' the Schrödinger equation using somewhat less rigorous methods. We will examine one of these methods here.

Schrödinger's first wave equation was actually a relativistic one, although when he finally published his work, he chose to present his derivation of the non-relativistic version. As we will see at the end of this chapter, the correct combination of quantum theory and Einstein's special theory of relativity gives rise to a new property of particle spin. Schrödinger's relativistic wave equation did not give this property but it is, none the less, a perfectly acceptable equation for quantum particles with zero spin.

It is possible to follow Schrödinger's reasoning from notebooks he kept at the time. His starting point was the well known equation of classical wave motion, which interrelates the space and time dependences of the waves. This equation can be separated into two further equations, one dealing only with the spatial variations of the waves, the other

---

Fig. 1.3 The buildup of an electron interference pattern. In photograph (a), the passage of 10 electrons through a double-slit apparatus has been recorded. In (b)–(e) the numbers recorded are 100, 3000, 20 000 and 70 000 respectively. (Reprinted with permission from Tonomura, A., Endo, J., Matsuda, T., and Kawasaki, T. (1989). *American Journal of Physics*, **57**, 117–20.)

† Quotation from Bloch, Felix (1976). *Physics Today*. **29**, 23.

dealing only with their time dependence. For waves oscillating in three dimensions, the spatial wave equation takes the form

$$\nabla^2\psi = -k^2\psi, \tag{1.17}$$

where ∇^2, the Laplacian operator, is given by $\partial^2/\partial x^2 + \partial^2/\partial y^2 + \partial^2/\partial z^2$ and k, the wave vector, is equal to $2\pi/\lambda$ where λ is the wavelength. There is a whole range of functions ψ (called wavefunctions) that satisfy this equation, ranging from simple sine and cosine functions to more complicated functions.

Now, according to de Broglie, $\lambda = h/p$, where p is the linear momentum of a wave–particle. If we make the non-relativistic assumption that $p = mv$, where m is equal to m_0, the rest mass of the particle (contrast this with eqn (1.13)), and v is its velocity, we can write

$$k = \frac{2\pi}{\lambda} = \frac{2\pi p}{h} = \frac{2\pi mv}{h}, \tag{1.18}$$

and hence

$$\nabla^2\psi = -\frac{4\pi^2 m^2 v^2}{h^2}\psi. \tag{1.19}$$

The total energy of a particle E is the sum of its kinetic and potential energies, i.e.

$$E = \frac{1}{2}mv^2 + V \tag{1.20}$$

where V is the potential energy. This expression can be rearranged to give

$$mv^2 = 2(E - V) \tag{1.21}$$

which, when inserted into eqn (1.19), yields

$$\nabla^2\psi = -\frac{8\pi^2 m}{h^2}(E - V)\psi \tag{1.22}$$

or

$$-\frac{\hbar^2}{2m}\nabla^2\psi + V\psi = E\psi \tag{1.23}$$

where $\hbar = h/2\pi$. This is the three-dimensional Schrödinger wave equation.

Simple isn't it? This 'derivation' probably follows Schrödinger's original quite closely. However, the reasoning behind it is almost too simplistic and when he came to publish his results Schrödinger elected to present a much more obscure derivation, one which did not refer either to the de Broglie hypothesis or the quantization of energy. In fact, all that he had done was to take the well known equation of classical wave motion and substitute for the wavelength according to de Broglie's relation. This in itself is perhaps not so remarkable; it was what Schrödinger did next that changed the world of physics for good.

The hydrogen atom

Schrödinger presented his wave mechanics to the world in a paper he submitted to the journal *Annalen der Physik* towards the end of January 1926, barely three weeks after he had made his initial discovery. In this paper he not only offered his (somewhat obscure) 'derivation' of the wave equation, but also applied the new theory to the hydrogen atom. It was this first application of wave mechanics that caught the attention of the physics community. Had he simply presented the wave equation, perhaps few physicists would have been convinced of its significance.

The earlier Rutherford–Bohr model of the hydrogen atom is essentially a planetary model, consisting of a massive central nucleus, the proton, orbited by a much lighter electron. The potential energy of the nucleus is spherically symmetric, and so a more logical coordinate system for the problem is one of spherical polar coordinates rather than traditional Cartesian (x, y, z) coordinates. Transformation of eqn (1.23) to a polar coordinate system produces quite a complicated differential equation and, although Schrödinger was an accomplished mathematician, he needed help to solve it. However, assistance was at hand in the form of a colleague at Zürich, Hermann Weyl.

Schrödinger's aim was to show that the quantum numbers introduced in a rather *ad hoc* fashion by Bohr emerged 'in the same natural way as the integers specifying the number of nodes in a vibrating string'.[†] This refers to the pictures, familiar to every undergraduate scientist, of standing waves generated in a string which is secured at both ends. A variety of standing waves are possible provided they meet the requirement that they 'fit' between the string's secured ends, i.e. they must contain an integral number of half-wavelengths. Thus, the longest frequency standing wave is characterized by a wavelength which is twice the length of the string (no nodes). The next wave is characterized by a wavelength equal to the length of the string (one node), and so on. The problem is more difficult for the hydrogen atom since now we are dealing with three-dimensional standing waves confined by a spherical potential, but the principles are the same.

[†] Schrödinger, E. (1926). *Annalen der Physik*. **79**, 361.

In order to obtain 'sensible' solutions of the wave equation for the hydrogen atom, it is necessary to restrict the range of functions that we will admit as acceptable. In particular, the acceptable functions must be single valued (only one value for a given set of coordinates), finite (no infinities) and continuous (no sudden 'breaks' in the functions). The last requirement must be met because the wave equation is a second-order differential equation, and a discontinuous function has no second differential.

Imposing these conditions on the wavefunctions is all that is necessary to produce the quantum numbers. Schrödinger wrote:[†] 'What seems to me to be important is that the mysterious "whole number requirement" no longer appears, but is, so to speak, traced back to an earlier stage: it has its basis in the requirement that a certain spatial function be finite and single-valued.' Thus, the integer numbers that appeared as if by magic in Bohr's theory of the atom are generated naturally in Schrödinger's. These integer numbers, the quantum numbers, are an intrinsic part of the acceptable solutions of Schrödinger's wave equation and hence, also, of the energies associated with these functions. The quantization of energy therefore follows from the standing wave condition applied to the electron in an atom.

We might add here that the differential equations we have been dealing with have a special property: a differential operator operates on a function to yield the same function multiplied by some quantity (in this case the energy E). The functions satisfying such equations are given the special name *eigenfunctions*, and the quantities are called the *eigenvalues*. Thus, when Schrödinger published his first paper on his new wave mechanics in 1926, its title was 'Quantization as an eigenvalue problem'.

The results that Schrödinger obtained for the wavefunctions of the hydrogen atom are familiar to every undergraduate scientist who has taken an introductory course in quantum mechanics. They are the electron orbitals and their three-dimensional shapes alone — which depend on the 'azimuthal' quantum number l and the 'magnetic' quantum number m_l — explain a great deal of chemistry. Their energies depend only on the principal quantum number n and are given by the same expression deduced by Bohr (eqn (1.8)).

Schrödinger's interpretation of the wavefunctions

Schrödinger's application of his new wave mechanics to the hydrogen atom was hailed as a triumph. However, although the new theory

---

[†] Schrödinger, E. (1926). *Annalen der Physik*. 79, 361.

explained the rules of quantization, it had merely shifted the burden of explanation from those rules to the wavefunctions themselves. A real understanding of the behaviour of sub-atomic particles, encompassing the full details of the relationship between the mechanics and the underlying physical reality, could only come through an interpretation of the wavefunctions. What were they?

In his first few papers on wave mechanics, Schrödinger referred to the wavefunction as a 'mechanical field scalar', a suitably obscure title for a function whose meaning was far from clear. Schrödinger was in fact convinced that the underlying reality was undulatory — that quantum theory was essentially a wave (or, more correctly, a field) theory. Thus, he initially interpreted the wavefunction as representing a vibration in an electromagnetic field, 'to which we can ascribe more than today's doubtful reality of the electronic orbits'.[†]

Schrödinger supposed that transitions between standing waves representing the stationary quantum states of an atom are smooth and continuous. He was hopeful that he could explain the apparent non-classical properties of atoms with essentially classical concepts, and thereby recover some of the cherished notions of determinism and causality that quantum theory seemed to abandon.

He therefore viewed an atomic electron not as a particle, but as a collection of wave disturbances in an electromagnetic field. He proposed that the electron's particle-like properties are really manifestations of their purely wave nature. When a collection of waves with different amplitudes, phases and frequencies are superimposed, it is possible that they may add up to give a large resultant in a specific region of space (see Fig. 1.4). Such a superposition of waves is commonly called a wave 'packet'. Schrödinger argued that, since the square of the amplitude of the resultant is related to the strength of the field as a function of position, the movement of a wave packet through space might, therefore, resemble the movement of a particle. This is in many ways analogous to the relationship between geometrical (ray) optics and wave optics. According to this view, the dual wave–particle nature of sub-atomic particles is replaced by a purely wave interpretation, with the wavefunctions representing the amplitudes of a field.

This explanation is not entirely satisfactory, as Hendrik Lorentz pointed out in a letter to Schrödinger. When confined to move in a small region of space, such a wave packet is expected to spread out rapidly, dispersing or 'dissolving' into a more uniform distribution. This is obviously not what happens to sub-atomic particles like electrons.

Schrödinger had other problems too. He did not like the fact that

[†] Schrödinger, E. (1926). *Annalen der Physik*. **79**, 361.

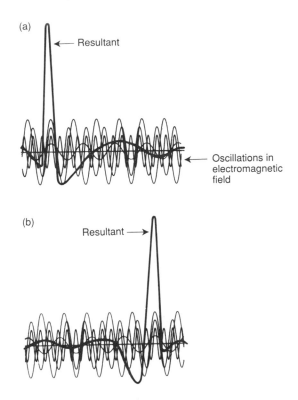

Fig. 1.4 The motion of a wave packet. (a) The amplitudes, phases and frequencies of a collection of waves combine and constructively interfere to form a resultant wave packet with a large amplitude confined to a specific region of space. (b) As all the individual waves move, so too does the region of constructive interference.

the wavefunctions could be complex (i.e. contain $\sqrt{-1}$), preferring to believe that any description of a microphysical reality worth its salt ought to involve 'real' functions. (In fact, the presence of complex terms in the wavefunction provides the all-important phase information responsible for interference effects.)

In addition, the wavefunctions for complicated systems containing two or more particles are functions not just of three spatial coordinates, but of many coordinates. In fact, the wavefunction of a system containing $3N$ particles depends on $3N$ position coordinates and is a function in a $3N$-dimensional 'configuration space'. This might be all very well for a mathematician, but remember that Schrödinger was looking for the reality that was supposed to lie beneath his wave mechanics: it is difficult to visualize a reality in an abstract, multi-dimensional space. Furthermore, we are quite free to choose the kind of space in which to represent the wavefunction. 'Momentum space', in which the momenta of the particles serve as coordinates, is just as acceptable mathematically as position space, and yet the wavefunctions in these two representations look distinctly different.

To a certain extent, these problems of interpretation were clarified by Max Born. But Born's interpretation of the wavefunctions was not to everyone's taste: Schrödinger and Einstein in particular did not like it one bit.

Born's probabilistic interpretation

Max Born wrote a short paper about the quantum mechanics of collisions between particles which was published in 1926 at about the same time as Schrödinger's fourth paper in the series 'Quantization as an eigenvalue problem'. Born rejected Schrödinger's wave field approach. He had been influenced by a suggestion made by Einstein that, for photons, the wave field acts as a strange kind of 'phantom' field, 'guiding' the photon particles on paths which could therefore be determined by wave interference effects. Thus, reasoned Born, the square of the amplitude of the wavefunction in some specific region of configuration space is related to the *probability* of finding the associated quantum particle in that region of configuration space. This automatically leads to the concept of normalization, which we will discuss in the next chapter.

At first sight, Born's interpretation seems unremarkable. After all, we know that the square of the amplitude of a light 'wave' in a specific region of space is related to its intensity and, from the photoelectric effect, we know the intensity is in turn related to the number of photons present in that same region of space. However, Born's way of thinking represented a marked break with classical physics. Unlike Schrödinger, who wanted to invest an element of physical reality in the wavefunctions, Born argued that they actually represent our *knowledge* of the state of a physical object.

Born's interpretation solved many of the problems raised by Schrödinger's wave mechanics. According to Born, the wavefunctions are not 'real' (in the sense that water waves are real) and so it does not matter that they are sometimes complex. The probability densities *must* be real, since they refer to measureable properties of quantum particles. Likewise, the tendency for Schrödinger's wave packet to spread out is a problem only if the wavefunctions are physically real. No such problem arises if the wavefunctions represent the evolution of our state of knowledge of a quantum system.

Born's interpretation was also consistent in some respects with the view being developed by Bohr and (as we will see below) Werner Heisenberg. Born argued that wave mechanics tells us nothing about the state of two quantum particles (such as electrons) following a collision: we can use the theory only to obtain probabilities for the various possible states. Just as Einstein had discovered for spontaneous transitions, quantum theory appeared to cut the direct link between cause and effect.

In a later paper, Born showed that his interpretation allowed the calculation of the probability for a quantum transition between stable states (such as an atomic absorption or emission between electron orbitals). As we saw in Section 1.2, the model of a classical electron moving between stable atomic orbits fails because such an electron would be expected to radiate energy during the transition. Born argued that Schrödinger's interpretation in terms of vibrations in an electromagnetic field did not help, since it could not be used to explain how an electron removed completely (i.e. ionized) from a stable atomic orbit could produce a discrete track in an ionization chamber. He therefore combined wave mechanics with the idea of quantum jumps implied by Bohr's theory of the atom. Born admitted that such a quantum jump, 'can hardly be described within the conceptual framework of Bohr's theory, nay, probably in no language which lends itself to visualizability.'[†] Cause and effect was once again threatened by instantaneous quantum jumps.

This, of course, was exactly what Schrödinger had been hoping to avoid. The purpose of his wave mechanics was to reintroduce a classical interpretation for the mechanics of the atom, albeit one of waves rather than particles. To add quantum jumps to this picture simply added insult to injury. In a heated debate between Schrödinger, Bohr and Heisenberg on the interpretation of quantum theory, an exasperated Schrödinger pleaded with an unyielding Bohr:[‡]

You surely must understand, Bohr, that the whole idea of quantum jumps necessarily leads to nonsense . . . If we are still going to have to put up with these damn quantum jumps, I am sorry that I ever had anything to do with quantum theory.

Cause and effect are part of our everyday lives, and are not things to be given up lightly. Many physicists found Born's interpretation unpalatable. Ironically, Born claimed that he had been influenced by Einstein, and yet in December 1926 Einstein wrote a letter to Born which contains

† Born, M. (1926). *Zeitschrift für Physik*. **40**, 167.
‡ Quotation from Heisenberg, Werner. (1969). *The part and the whole*. Verlag, Munich.

the phrase that has since become symbolic of Einstein's lasting dislike of the element of chance implied by quantum theory:[†]

Quantum mechanics is very impressive. But an inner voice tells me that it is not yet the real thing. The theory produces a good deal but hardly brings us closer to the secret of the Old One. I am at all events convinced that *He* does not play dice.

1.5 MATRIX MECHANICS AND THE UNCERTAINTY PRINCIPLE

Matrix Mechanics

Before leaving Göttingen to join Niels Bohr in Copenhagen, Werner Heisenberg developed a completely novel approach to quantum theory which became known as matrix mechanics. In this theory, physical quantities are represented not by their values as in classical physics, but by sets of time-dependent complex numbers. Unlike Schrödinger, Heisenberg was not particularly concerned to find some underlying physical reality—he was simply after a framework through which connections could be made between physical quantities in ways that would fit the known facts. Heisenberg's theory was essentially a mathematical algorithm—plug in the right numbers in the right way and you get the right answer.

Max Born and Pascual Jordan recognized that Heisenberg's 'sets' were actually matrices. A matrix is an array of numbers which can take the form of a column, row, rectangle or square. Just as there are rules for combining ordinary numbers in addition, subtraction, multiplication and division, there are also rules for combining matrices. One very important consequence of the rule for matrix multiplication is that the result can sometimes depend on the order in which two matrices are multiplied together, i.e. it is possible that the product of two matrices **A** and **B** is not necessarily equal to the product of **B** and **A**. Matrices for which $AB \neq BA$ are said not to commute. Obviously, for ordinary numbers $AB = BA$ and so ordinary numbers always commute.

Born, Jordan and Heisenberg reformulated Heisenberg's original theory as matrix mechanics. This approach was very successful—it too explained many of the otherwise inexplicable features of quantum phenomena. In 1926, Wolfgang Pauli showed how the theory could be used to explain the hydrogen atom emission spectrum. But now phys-

[†] Einstein, Albert, letter to Born, Max, 4 December 1926.

icists had yet another problem to confront: matrix mechanics and wave mechanics were formulated and presented at about the same time (late 1925 to early 1926), and although the predictions of the theories were the same, they were quite different in approach. Which was right?

To a certain extent, the answer to this question was provided by Schrödinger in a paper published in 1926. He demonstrated that matrix mechanics and wave mechanics give the same results because the two theories are mathematically equivalent: they represent two different ways of addressing the same problem. Of course, Schrödinger argued, wave mechanics is to be preferred because it offers a conceptual basis for understanding the behaviour of quantum particles which matrix mechanics could never have. Many physicists tended to agree, although some dissented. For example, Lorentz confessed to a preference for matrix mechanics because of the problems with Schrödinger's wave packet idea. However, all physicists were a little uneasy in the knowledge that a theory as important as quantum theory could be expressed in two totally different ways.

The real connection between matrix mechanics and wave mechanics was made clear by the mathematician John von Neumann in the early 1930s. He showed that wave mechanics could be expressed in an operator algebra. We will see in Chapter 2 that, where matrix mechanics depends upon the properties of non-commuting matrices, wave mechanics can be derived from the properties of non-commuting operators.

Heisenberg's uncertainty principle

Any introductory course on quantum mechanics will contain an early discussion of Heisenberg's famous uncertainty principle. Unfortunately, like most of the material we have covered so far in this chapter, the uncertainty principle is often presented to modern undergraduate students in a matter-of-fact way. Students are told what it is, how it fits into the structure of quantum theory and how it applies to physical systems. It all seems very neat but, in fact, the uncertainty principle was formulated in the midst of argument about the interpretation of quantum theory. Despite the fact that today we know quite a bit more about the theory and its applications, arguments about the meaning of the uncertainty principle are no less heated and confusing now than they were in 1926.

Towards the end of that year, it was clear that Schrödinger's views were winning out: many physicists who expressed a preference opted for wave mechanics because it appeared to offer the best prospects for further interpretation. Bohr and Heisenberg tried hard to persuade Schrödinger of the importance of the idea of quantum jumps, as we have seen, but failed. Bohr and Heisenberg did not give up, however. They became more determined than ever to resolve the difficulties of inter-pretation by taking a radical approach.

The problem with matrix mechanics was its abstract nature. Whereas Born's probabilistic interpretation of Schrödinger's wavefunction seemed to be at least consistent with the idea of an electron path or trajectory, no such trajectory is defined in matrix mechanics. But then, Schrödinger's own interpretation of his wave mechanics was self-contradictory: the motion of a wave packet could not be used to describe the path of an electron because of the tendency of the wave packet to disperse. Anyone who had looked at the track left by an electron in a cloud chamber could be convinced of the reality of the electron's particle-like properties and yet this was something that Schrödinger's interpreta-tion appeared unable to rationalize.

The situation was very confusing. It was at this point that Bohr and Heisenberg decided to go right back to the drawing board. They began to ask themselves some fairly searching, fundamental questions, such as: What do we actually mean when we speak about the *position* of an electron? The track caused by the passage of an electron through a cloud chamber seems real enough — surely it provides an unambiguous measure of the electron's position? But wait: the track is made visible by the con-densation of water droplets around atoms that have been ionized by the electron. This process of ionization is a quantum process and therefore subject to the rules, and open to the probabilistic interpretation, of quan-tum mechanics. According to this interpretation, it is the large number of probabilistic (and hence indeterminate) ionizations which allows what seems to be a classical, deterministic path to be made visible.

In 1927, Heisenberg decided that to talk about the position and momentum of any object requires an operational definition in terms of some experiment designed to measure these quantities. To illustrate his reasoning, Heisenberg developed a 'thought' experiment involving a hypothetical γ-ray microscope. Supposing we wished to measure the path of an electron — its position and velocity (or momentum) as it travels through space. The most direct way of doing this would be to follow the electron's motion using a microscope. Now the resolving power of an optical microscope increases with increasing frequency of radiation, and so a γ-ray microscope would be necessary to give the spatial resolution required to 'see' an electron. The γ-ray photons bounce off the electron, some are collected and produce the magnified image.

But we have a problem. γ-rays consist of 'big', high-energy photons (remember $\varepsilon = h\nu$) and, as we know from the Compton effect, each time a γ-ray photon bounces off an electron, the electron is given a severe jolt. This jolt means that the direction of motion and the momentum of the electron are changed in ways that are unpredictable. According to Born's interpretation of the wavefunction of the electron, only the probabilities for scattering in certain directions and with certain momenta can be calculated using quantum mechanics. Although we might be able to obtain a fix on the electron's instantaneous position, the sizeable interaction of the electron with the device we are using to measure its position means that we can say nothing at all about the electron's momentum.

We could use much lower energy photons in an attempt to avoid this problem and so measure the electron's momentum, but we must then give up hope of determining its position. Heisenberg reasoned that the exact position and momentum of a quantum particle could not be measured simultaneously. To determine these quantities requires two quite different kinds of measuring apparatus and the measurement of one excludes the simultaneous measurement of the other.

Heisenberg used Born's probabilistic interpretation of the wavefunction to derive an expression for the 'uncertainties' (actually root-mean-square deviations) of the position and linear momentum of a particle confined to move in one dimension (along the x coordinate). He obtained

$$\Delta x \Delta p_x \geqslant h/4\pi. \tag{1.24}$$

Thus, fixing the position of an electron exactly ($\Delta x = 0$) implies, from $\Delta p_x \geqslant h/4\pi\Delta x$, an infinite uncertainty in the electron's momentum, and vice versa. Extending these arguments to the measurement of energy, Heisenberg obtained

$$\Delta E \Delta t \geqslant h/2\pi. \tag{1.25}$$

This expression is often presented as an energy–time 'uncertainty' relation, but is really a reworking of the position–momentum uncertainty relation in the context of the time-dependent Schrödinger equation (which we will meet briefly in the next chapter). The relation (1.25) is usually interpreted in a practical sense to signify that the moment of emission (say) of a quantum particle will be uncertain by an amount Δt related to the uncertainty in its energy. The more sharply we can measure (in time) the creation or passage of a quantum particle, the more uncertain will be its energy, and vice versa.

Interpretation of the uncertainty principle

Some physicists have argued that the uncertainty principle represents the starting point from which the whole of quantum mechanics can be deduced. It is apparent that Heisenberg himself thought something along these lines. He did not believe that it was necessary to use terms like 'wave' or 'particle' when talking about quantum phenomena and preferred, instead, to continue with the supposition that the theory merely provided a 'consistent mathematical scheme [that] tells us everything which can be observed.' He concluded that 'Nothing is in nature which cannot be described by this mathematical scheme.' This is a purely 'instrumentalist' approach—the theory (in particular, the uncertainty principle) tells us that there are limits on what is *measureable* and it is impossible to do anything other than speculate on what is not measureable.

Bohr disagreed strongly with Heisenberg on this point. For him, it was wave–particle duality that lay at the heart of quantum mechanics. All the rest—including the uncertainty principle—were the physical and mathematical consequences of using two diametrically opposed classical concepts, waves and particles, to describe something that was fundamentally non-classical. According to Bohr, quantum theory tells us not what is measureable, but what is *knowable*.

Thus, according to this line of reasoning, Heisenberg's thought experiment involving the γ-ray microscope is flawed because it presupposes that a definite position and momentum can be defined for the electron: it is the act of measurement that makes their joint specification impossible. For Bohr, the uncertainty principle indicates that the very idea (one might say the physical reality) of the electron's position and momentum are undefined until the act of measurement.

Bohr put Heisenberg under intolerable pressure—so much so that at one point Heisenberg was reduced to tears. They finally managed to reach a compromise and, in the paper in which Heisenberg presented his uncertainty principle, he added a footnote containing the sentence: 'I am greatly indebted to Professor Bohr for having had the opportunity of seeing and discussing his new investigations which are soon to be published as an essay on the conceptual structure of quantum theory.' This was to be Bohr's notion of *complementarity*, discussed in detail in Chapter 3.

Heisenberg has summarized this period of intense debate as follows:[†]

I remember discussions with Bohr which went through many hours till very late at night and ended almost in despair; and when at the end of the discussion I went alone for a walk in the neighbouring park I repeated to myself again and again the question: Can nature possibly be as absurd as it seemed . . .?

[†] Heisenberg, Werner (1989). *Physics and philosophy*. Penguin, London.

1.6 RELATIVITY AND SPIN

The years 1925–27 saw much of the structure of quantum theory set in place. Although many significant developments have happened since, and arguments about the interpretation of the theory still abound, the mathematical formalism of the non-relativistic version of the theory has remained more-or-less unchanged. There were still a few problems, however, and solving them led the English physicist Paul Dirac to some spectacular conclusions about the nature of matter.

Electron spin

Bohr's theory of the atom did a fine job of explaining the absorption and emission spectra of one-electron atoms in terms of quantum rules. Schrödinger's wave mechanics did better in the sense that the rules of quantization were given a firm mathematical basis. However, experimental spectra revealed quite a number of problems for which wave mechanics did not appear to have solutions. In particular, some atomic emission lines predicted by the theory were seen to be split in the presence of a magnetic field into two quite distinct lines in experimental spectra. Wave mechanics could not explain this phenomenon.

In 1925, Samuel Goudsmit and George Uhlenbeck proposed that the splitting of these lines arises because the electron in an atom responsible for the emission transition possesses an intrinsic angular momentum – it is spinning on its axis in much the same way that the earth spins on its axis as it orbits the sun. A spinning electric charge moving in an electromagnetic field generates a small, local, magnetic field. The spin magnetic moment of the electron can become aligned with or against the lines of force of an applied magnetic field, giving two states of different energy. In the absence of this splitting, there would be only one state and hence only one line in the atomic emission spectrum. Instead, the interaction produces two distinct states, giving rise to two emission transitions.

While this is a very picturesque interpretation, its problems become apparent as soon as we abandon Bohr's planetary model of the atom in favour of Schrödinger's wave mechanics. The appearance of only two lines means that the electron cannot acquire just any old angular momentum. The angular momentum intrinsic to an electron is not only quantized (only certain values are allowed), it is restricted to only two possible values. This contrasts with the quantization associated with the principal and azimuthal quantum numbers. And where was electron spin in Schrödinger's wave mechanics of the hydrogen atom?

The Dirac equation

At the end of 1926, Heisenberg and Paul Dirac agreed to a bet about how soon spin could be understood within the framework of quantum theory. Heisenberg suggested three years, Dirac three months. Neither was exactly on the mark, but Dirac was closer. On 2 January 1928, Dirac himself submitted a paper to the *Proceedings of the Royal Society* which set out the correct relativistic quantum theory of the electron, from which electron spin emerged naturally.

Einstein's special theory is in many ways all about the correct treatment of time as a kind of fourth dimension, on an equal footing with the three conventional spatial dimensions, x, y and z. In fact, it is ct which constitutes a fourth dimension (note that it has the same units of length as the other three). Schrödinger's original time-dependent wave equation (which we will discuss briefly in Chapter 2) is 'unbalanced' in this regard, being a second-order differential equation in the three Cartesian coordinates but only a first-order differential equation in time.

Dirac derived a version of the wave equation for a free electron in which space and time are treated on an equal footing. This equation has some interesting features. It admits twice as many solutions for the wavefunctions as we might expect, half of them corresponding to states of negative energy. This is an inevitable consequence of using the correct relativistic expression for the energy of a freely moving particle, eqn (1.12), which is a quadratic equation. Dirac took these negative-energy solutions seriously, and went on to predict the existence of antimatter.

Furthermore, for Dirac's wave equation to make any sense it became clear that the wavefunctions had to take the form of matrices. Half of each matrix refers to states of the electron and the other half to states of the positron, the antiparticle of the electron. If we consider only those solutions with positive values for the energy, the wavefunctions are two-component *spinors* — 2×1 or 1×2 matrices. Dirac was able to show that these components are equivalent to two possible orientations of the electron's magnetic moment: they represent the two spin orientations of an electron.

Now the introduction of a four-dimensional space–time results in the need for a fourth 'degree of freedom' for the electron in addition to the three degrees of freedom corresponding to translation in the x, y and z directions. That fourth degree of freedom requires the specification of a fourth quantum number, usually given the symbol s, which according to the theory can take only the value $\frac{1}{2}$.

Whatever it is, the property of electron spin does not correspond in any way to the notion of an electron spinning on its axis. Here we see the first example of a purely quantum property—electron spin has no counterpart in classical physics. To see why this is so, it is necessary to look at how classical properties can be derived from quantum properties in the limit that h tends to zero. The angular momentum of an electron orbiting a nucleus is related to the azimuthal quantum number l and there is no restriction on the size of l. Thus, the tendency for orbital angular momentum to disappear as h tends to zero can be compensated by increasing the size of l to infinity. The result can be non-zero, and so orbital angular momentum has a clearly defined classical counterpart (as indeed we should expect). However, the same is not true of electron spin. The spin quantum number s is a fixed quantity ($s = \frac{1}{2}$), and so cannot be increased to infinity as h tends to zero. The property analogous to electron spin therefore disappears for classical objects.

Although its interpretation is obscure, we do know that electron spin produces effects which give rise to a small magnetic moment. This moment can become aligned in the direction of an applied magnetic field or against that direction. We have learned to think of these two possibilities as 'spin-up' and 'spin-down'. In a magnetic field, the two possible orientations of the electron's magnetic moment give rise to two energy levels which are characterized by the magnetic quantum numbers $m_s = +\frac{1}{2}$ and $m_s = -\frac{1}{2}$. These quantum numbers correspond to the spin-up and spin-down states of the electron, and the two levels give rise to two lines in an atomic emission spectrum.

Quantum field theory

The modern form of relativistic quantum theory is called quantum field theory. In this theory, the spatial extension of a quantum particle due to its wave nature is recognized by its representation as a quantum field. This is more than Schrödinger's simple wave field idea, although there are obvious similarities. For example, an electron wavefunction is thought of as a specific excitation (vibration) of an electron field, and is interpreted as a probability amplitude just as in ordinary quantum mechanics.

Quantum electrodynamics—the quantum field version of Maxwell's classical electrodynamics—deals with the forces of electromagnetism. This theory has proved to be tremendously powerful and successful, but it has done nothing to dispel the difficulties over interpretation. Although the mathematics of quantum theory has developed and has become more sophisticated since it was first formulated over 60 years ago, the problems of interpretation remain. We are still left with the uncertainty principle, the idea of quantum jumps and the wavefunctions.

We are still left to decide whether we must abandon direct cause-and-effect. The progress that has been made in the last 60 years has certainly improved the predictive power of the theory, but it has really been a matter of sharpening the mathematical formalism rather than our understanding of it.

2
Putting it into practice

2.1 OPERATORS IN QUANTUM MECHANICS

Now that we have satisfied ourselves of the need for a quantum theory of radiation and matter, and we have seen the different positions taken up by the major figures in the theory's early development, we must interrupt our historical narrative and turn our attention to details of how the theory works in practice. Readers are reminded that this is not a textbook, and so the description we will give here is a highly selective one.

Firstly, it is important to see how the modern version of the theory is constructed from a set of postulates — statements accepted to be true without proof which are justified later through agreement with experiment. We will frequently return to the philosophical implications of such an approach. Secondly, in order to appreciate the details of the debate between Bohr and Einstein, and further arguments presented by Einstein and others which led to the development of important experimental test cases, it is necessary to know how the theory is routinely applied. Finally, many of these crucial experimental tests have been performed by exploiting the properties of pairs of photons with correlated polarizations, and so we have included here discussions of the Pauli principle (important for any subsequent discussion of the behaviour of two-particle quantum systems), together with a section on the polarization properties of photons. These properties can be used to illustrate in a simple way the problems of quantum measurement. All this material should set us up nicely for the remainder of the book.

Much of the material covered in this chapter is tied up with the mathematical formalism of quantum theory. Readers completely unfamiliar with this formalism may therefore find this material somewhat hard to digest. The unfamiliar becomes familiar with experience and practice, and is in this case well worth the effort required. You will soon discover that although the formalism of quantum theory presented here may appear complicated, the complications often arise in the language used, not in the algebra itself.

Before we begin, it is important to emphasize that the quantum theory described in this chapter is essentially that taught widely to modern undergraduate students of chemistry and physics. That this theory is the best description of the microphysical world of elementary particles, atoms and molecules is not in dispute. It is what the theory means in terms of the relationship between its concepts and physical reality that has caused (and continues to cause) so much argument and debate. Thus, in the account which follows, we should note that the concepts of the theory are given an interpretation that is not readily accepted by every scientist.

Mathematical operators

Despite their mathematical equivalence, the quantum theory taught to modern undergraduate students of chemistry and physics is based mostly on Schrödinger's wave mechanics rather than Heisenberg, Born and Jordan's matrix mechanics. This is because students learn about the algebra of ordinary functions and their associated mathematical operators first: the mathematics of matrices is often regarded as a topic for more advanced courses. Consequently, the mathematics of the operator (Schrödinger) form of quantum mechanics is more familiar (although this does not necessarily mean that it is easy to follow). An experienced, professional quantum 'mechanic' will quite happily move between both descriptions.

The most common mathematical language of quantum theory is therefore that of operator algebra, and the postulates of the theory are most often presented to undergraduates in this language. Operators are very familiar to us; they are simply instructions to do something to a function — multiply it, differentiate it etc. However, it takes a little while to get used to the idea of handling the operators by themselves, i.e. without having the functions present in an equation. In fact, in an equation containing only mathematical operators, the existence of some function in the appropriate coordinates on which the operators are supposed to operate is implicitly assumed. We will discover later in this chapter that we can go quite a long way in our analysis of quantum systems by considering the properties of operators and the assumed properties of the wavefunctions without actually having to solve the appropriate Schrödinger equation.

The position–momentum commutation relation

Suppose I throw an apple into the air and you photograph it as it falls to the ground using a camera fitted with a rapid autowind facility. You take a sequence of photographs at fixed intervals in time as the apple falls to the ground. Each photograph is a record of the position of the apple (its height x above the ground) at a particular time in its motion. We could analyse the sequence of photographs to find the values of the apple's position and velocity (and hence its momentum in the x coordinate, p_x) as a function of time.

Further suppose that, for some obscure reason, we need to calculate the product of the apple's position and momentum. I choose to determine the product by multiplying position by momentum. You choose to multiply momentum by position. No matter, since we know that for ordinary numbers and their associated units, $xp_x = p_xx$, or

$$xp_x - p_xx = 0 \tag{2.1}$$

and so we expect to get the same answers. Equation (2.1) is generally called a commutation relation, sometimes abbreviated using the notation $[x, p_x]$.

This might seem trivial, but the results of experiments lead us to conclude that eqn (2.1) does *not* hold for quantum particles like electrons or photons. The corresponding quantum mechanical version of eqn (2.1) is

$$xp_x - p_xx = i\hbar \tag{2.2}$$

where $i = \sqrt{-1}$. This is known as the quantum mechanical position–momentum commutation relation. As we have said before, Planck's constant h is a very small quantity and so measurements made on macroscopic objects like apples will never reveal behaviour other than that described by eqn (2.1). However, for microscopic objects like electrons, the magnitude of Planck's constant becomes extremely important. Incidentally, eqn (2.2) demonstrates once again that classical mechanics can be recovered from quantum mechanics by approximating h to zero.

We have at least two ways of proceeding from here. We cannot explain eqn (2.2) if we treat x and p_x as ordinary quantities. Either we treat them as non-commuting matrices (cf. Section 1.4) or as non-commuting mathematical operators. If we choose to use matrices, we are led to matrix mechanics. If we choose operators, we have to assume that there must exist some function which depends on x (let us call it a wavefunction ψ) on which the operators are supposed to operate. Putting this wavefunction into eqn (2.2) gives

$$(\hat{x}\hat{p} - \hat{p}\hat{x})\psi = \hat{x}\hat{p}\psi - \hat{p}\hat{x}\psi = i\hbar\psi \tag{2.3}$$

where we have used carets ($\hat{}$), to remind us that $\hat{x}$ and $\hat{p}_x$ should now be regarded as mathematical operators.

Now we have to decide what form $\hat{x}$ and $\hat{p}_x$ should take to satisfy eqn (2.3). We can start by making the operator $\hat{x}$ equivalent to 'multiplication by the value of x', just as in classical mechanics and setting $\hat{p}_x$ proportional to $\partial/\partial x$ (for reasons that will become immediately apparent). We use a partial differential operator because we need to assume that the values of any other coordinates on which ψ might depend are kept constant for the purposes of differentiation. Let us in fact assume that $\hat{p}_x = a\partial/\partial x$, and use the commutation relation to find out what the constant a must be. From eqn (2.3) we have

$$xa\frac{\partial}{\partial x}\psi - a\frac{\partial}{\partial x}(x\psi) = i\hbar\psi. \tag{2.4}$$

Note that we have dropped the use of the caret on x because it is now defined as multiplication by a quantity rather than a more elaborate operator. This is a convention which we will follow throughout this book.

The second term on the left-hand side of eqn (2.4) is the differential of the product of two functions which depend on x and which we can expand using the product rule, $\partial(uv)/\partial x = v\partial u/\partial x + u\partial v/\partial x$:

$$xa\frac{\partial}{\partial x}\psi - a\left(\psi + x\frac{\partial}{\partial x}\psi\right) = i\hbar\psi. \tag{2.5}$$

When the bracket on the left-hand side of eqn (2.5) is expanded, the first and third terms cancel, $a = -i\hbar$ and so $\hat{p}_x = -i\hbar\partial/\partial x$. We can similarly define $\hat{p}_y = -i\hbar\partial/\partial y$ and $\hat{p}_z = -i\hbar\partial/\partial z$. The square of $\hat{p}_x$ is readily deduced from $\hat{p}_x^2 = \hat{p}_x.\hat{p}_x = -\hbar^2\partial^2/\partial x^2$ (remember $i^2 = -1$).

How to 'derive' the Schrödinger equation (again)

According to non-relativistic classical mechanics, the total energy of a particle is the sum of its kinetic and potential energies, eqn (1.20), which can be written in terms of $p = mv$ as follows:

$$\frac{p^2}{2m} + V = E. \tag{2.6}$$

To get the equivalent quantum mechanical expression, we replace the physical quantity p^2 with its quantum mechanical operator equivalent $\hat{p}^2$, and introduce the wavefunction ψ. We take $\hat{p}^2$ to be equal to $(\hat{p}_x^2 + \hat{p}_y^2 + \hat{p}_z^2) = (-\hbar^2\partial^2/\partial x^2 - \hbar^2\partial^2/\partial y^2 - \hbar^2\partial^2/\partial z^2) = -\hbar^2\nabla^2$. The result is

$$-\frac{\hbar^2}{2m}\nabla^2\psi + V\psi = E\psi \tag{2.7}$$

which is, of course, the non-relativistic three-dimensional Schrödinger wave equation. This exercise demonstrates that in the equations of quantum mechanics, the values of observable quantities like position and momentum are replaced by the mathematical operators which yield these values when they operate on the wavefunction.

Operators and the uncertainty principle

In 1929, the physicist Howard Robertson showed that the product of the standard deviations of two non-commuting operators $\hat{A}$ and $\hat{B}$ is given by

$$\Delta\hat{A}\Delta\hat{B} \geqslant \frac{1}{2} |\langle C \rangle| \tag{2.8}$$

where C is related to the commutator $[\hat{A}, \hat{B}]$, by $[\hat{A}, \hat{B}] = iC$, and $|\langle C \rangle|$ represents the modulus of the average value of C. This result is quite general, independent of the significance of $\hat{A}$ and $\hat{B}$ or their interpretation as operators for physical quantities.

Some physicists have argued that eqn (2.8) justifies the view that the Heisenberg uncertainty relation should be regarded as a fundamental law of nature. Since $\Delta x \Delta \hat{p}_x \geqslant h/4\pi$, for x and $\hat{p}_x$ the term equivalent to $|\langle C \rangle|$ in eqn (2.8) has the value $\hbar$. Hence $[x, \hat{p}_x] = i\hbar$. We used this position–momentum commutation relation to deduce the form of $\hat{p}$ needed for the quantum mechanical version of the equation for the total energy of a particle, and arrived at the Schrödinger equation. Thus, in principle, all of quantum mechanics can, via eqn (2.8), be deduced from the uncertainty relation.

2.2 THE POSTULATES OF QUANTUM MECHANICS

Contrary to popular belief, science is a very untidy discipline. This is seen to be true nowhere more than in the historical development of quantum mechanics. The revolutionary new theory that was to replace the classical mechanics of Galileo and Newton in the microphysical world was born amid confusion and desperation and grew up amid confusion and argument. By the end of the 1920s most physicists accepted that quantum theory had quite a lot going for it, but were unsure about how it should be interpreted. The theory had tremendous predictive power, and it scored success after success when put to the test in the laboratory. Where it failed the test, subtle but mathematically logical modifications to the theory were introduced which made it even more powerful. It became clear that, although the theory was difficult to understand, it was enormously useful and the gains from using it more than outweighed the doubts about what it meant.

Although the arguments about the interpretation of quantum theory were far from resolved, by the end of the 1920s there were few problems with its mathematical structure. Wave mechanics and matrix mechanics had been shown to be equivalent, and there could be no doubt that physicists knew *how* the theory should be applied. For some problems, wave mechanics offered the most accessible route to the solutions; for others matrix mechanics was the preferred choice.

We saw in Section 1.3 how Einstein, faced with the problem of explaining a fixed speed of light, raised that fact to the status of a postulate and used it to deduce the special theory of relativity. Schrödinger's version of quantum mechanics makes use of the mathematical properties of the wavefunctions and their relationships to observable quantities such as position, momentum, and energy. Although physicists might argue about the meaning of the wavefunctions, there was no dispute about how the wavefunctions should be manipulated to obtain results that could be compared with experimental measurements. Consequently, the existence of the wavefunctions and the sequence of operations that has to be done on them to obtain the results were elevated to the status of postulates. This was done principally by the mathematician John von Neumann, and the details were described in his book *Mathematical foundations of quantum mechanics*, first published in Berlin in 1932. The postulates promote the view that, rather than worry overmuch about where the wavefunctions come from, we might as well accept that they exist and use them to tell us interesting things about the microscopic world of molecules, atoms, electrons, and photons.

The postulates of quantum mechanics are the foundation stones upon which the most widely accepted version of the theory is built. From now on, we will refer to this most widely accepted version as the 'orthodox' interpretation of quantum theory. For the present, we will accept the postulates without question—to a certain extent, their justification comes from the fact that the theory which flows from them is arguably the most highly successful theory of physics ever devised. We will reserve our discussion of their validity until later chapters.

The wavefunction

Postulate 1. The state of a quantum mechanical system is completely described by the wavefunction ψ_n.

Here the subscript n serves as a shorthand for the set of one or more quantum numbers on which the wavefunction will depend. For example, the wavefunction of an electron in a hydrogen atom is completely specified by the set of quantum numbers n, l, m_l and m_s. The values $n = 2$, $l = 1$, $m_l = 0$ and $m_s = +\frac{1}{2}$ refer to one, and only one, state of the electron (recognizable, by convention, as a $2p_z$ electron with spin up). We will use ψ_n (or ψ_m, ψ_k, etc) to denote wavefunctions that refer to specific states of a quantum system and will continue to use ψ to denote a general wavefunction.

The wavefunction will be a mathematical function in whatever coordinate space is appropriate for the system under study, and there will be as many quantum numbers as there are coordinates. For example, electron wavefunctions for the hydrogen atom can be represented in three spherical polar coordinates and an 'internal' coordinate associated with the spin of the electron. For the present, we will leave open the exact nature of the wavefunction, but will accept that it somehow 'contains' all the information we need about the quantum system it describes.

According to Born's interpretation, the wavefunctions represent probability 'amplitudes', and the square of the wavefunction gives the probability density of the quantum particle in a particular region of space. Because the wavefunctions may be complex, it is necessary to calculate this probability density from the product $|\psi|^2 = \psi^*\psi$ if the result is going to be a real quantity (ψ^* is the complex conjugate of ψ). If we wish to know the probability of finding the particle in a specific region of space, it is necessary to multiply the probability density by the volume element to which we refer (much as we would determine the mass of some material by multiplying its mass density by its volume). Because the 'space' in which the wavefunction is represented may be a complicated, many-dimensional configuration space, we replace the volume element by a generalized spatial element $d\tau$. Thus, the probability of finding the particle in the region $d\tau$ is given by $\psi^*\psi d\tau$. Because there is only one particle and it must be found somewhere, the integral over all spatial elements $d\tau$ must be unity, i.e.:

$$\int_0^\infty \psi^*\psi\, d\tau = 1. \qquad (2.9)$$

This is known as the condition of normalization.

Operators for observables

Postulate 2. Observable quantities are represented by mathematical operators. These operators are chosen to be consistent with the position–momentum commutation relation.

The term observable quantity (or just 'observable') is used to signify all the quantities that we could, in principle, measure, such as position, momentum, energy etc. By choosing operators that are consistent with $[x, \hat{p}_x] = i\hbar$, we are taking the commutation relation (or, as we have seen, the uncertainty principle depending on your point of view) to be a fundamental 'truth' which is deemed not to be in need of proof. This is similar to Einstein's elevation of the speed of light to a fundamental constant whose invariance is accepted but cannot be proved.

According to this postulate, every observable must have a corresponding operator in the theory. We have seen in the previous section that the position operator can be chosen to be simply 'multiplication by x', as in classical mechanics. This choice for x forces the linear momentum operator $\hat{p}_x$ to take the form $-i\hbar\partial/\partial x$ in order to satisfy the commutation relation. The operator for the total energy is called the Hamiltonian operator $\hat{H}$. It consists of kinetic and potential energy operators, usually given the symbols $\hat{T}$ and V respectively. $\hat{T}$ is usually a differential operator (it is obviously closely related to $\hat{p}^2$) and V is usually just 'multiplication by V'. Thus, the Schrödinger equation, eqn (2.7), can be written succinctly as follows:

$$\hat{H}\psi = E\psi \qquad (2.10)$$

where

$$\hat{H} = \hat{T} + V = -\frac{\hbar^2}{2m}\nabla^2 + V. \qquad (2.11)$$

Equation (2.10) has a form characteristic of an eigenvalue equation: a mathematical operator (in this case $\hat{H}$) operates on some function to give the same function multiplied by the value of some quantity (in this case E). Functions that satisfy such an equation are said to be eigenfunctions of the operator ($\hat{H}$) and the corresponding values of the quantities (the energies) are called eigenvalues. We can say that if the specific wavefunction ψ_n is an eigenfunction of $\hat{H}$, then $\hat{H}\psi_n = E_n\psi_n$ and its eigenvalue is E_n. Different specific eigenfunctions $-\psi_k, \psi_m-$ may or may not have different energy eigenvalues $-E_k, E_m$. For example, the energies of the electron wavefunctions (or orbitals) of the hydrogen atom depend only on the principal quantum number n. In the absence of a magnetic field, states with the same value of n but different values of l, m_l or m_s will therefore have the same energy.

The values of observables

Postulate 3 The mean value of an observable is equal to the expectation value of its corresponding operator.

For a specific wavefunction ψ_n, the expectation value of some operator $\hat{A}$ is defined by the expression:

$$\langle A_n \rangle = \frac{\int \psi_n^* \hat{A} \psi_n \, d\tau}{\int \psi_n^* \psi_n \, d\tau}. \tag{2.12}$$

This expression appears to have arrived out of the blue, but it is, in fact, related to an equivalent expression from probability calculus. Let us examine how the expectation value is related to the value of an observable for the case when ψ_n is an eigenfunction of $\hat{A}$. Denoting the corresponding eigenvalue as a_n, we have $\hat{A}\psi_n = a_n \psi_n$. If we multiply both sides of this equation from the left by ψ_n^* and integrate over all spatial elements, we obtain:

$$\int \psi_n^* \hat{A} \psi_n \, d\tau = a_n \int \psi_n^* \psi_n \, d\tau. \tag{2.13}$$

Note that ψ_n^* on the left-hand side of eqn (2.13) multiplies the result of the operation of $\hat{A}$ on ψ_n and $\psi_n^* \hat{A} \psi_n \neq \hat{A} \psi_n^* \psi_n$. There is no such restriction on the right-hand side of eqn (2.13): a_n is just the value of some quantity which is independent of the coordinates and which can therefore be taken out of the integral. From eqn (2.13) we have

$$\langle A_n \rangle = \frac{\int \psi_n^* \hat{A} \psi_n \, d\tau}{\int \psi_n^* \psi_n \, d\tau} = a_n. \tag{2.14}$$

Equation (2.14) applies only if ψ_n is an eigenfunction of $\hat{A}$. Although postulate 3 refers to the 'mean value' of an observable, we can see from eqn (2.14) that when ψ_n is an eigenfunction of $\hat{A}$ the expectation value is exactly a_n. We will see later that the use of the word 'mean' becomes necessary when we consider functions which are not eigenfunctions. Obviously, if ψ_n is normalized, $\int \psi_n^* \psi_n \, d\tau = 1$ and $\langle A_n \rangle = \int \psi_n^* \hat{A} \psi_n \, d\tau = a_n$.

We have seen how the wavefunctions may sometimes be complex functions. The operators too can sometimes be complex (as in $\hat{p}_x = -i\hbar \partial/\partial x$). However, if postulate 3 is to make any sense, the eigenvalue of an operator representing an observable must be a real quantity if we are to interpret it as the value of the observable, since this is something that we can, in principle, measure in the laboratory. This is an important restriction. Operators whose eigenvalues are exclusively real are called hermitian operators. Their integrals have the property that

$$\int \psi_n^* A \psi_m \, d\tau = \left(\int \psi_m^* A \psi_n \, d\tau \right)^* \tag{2.15}$$

where ψ_m and ψ_n are eigenfunctions of $\hat{A}$.

Furthermore, as a result of this property of the operator (called hermiticity) any two eigenfunctions are also orthogonal, i.e.

$$\int \psi_n^* \psi_m \, d\tau = 0. \qquad (2.16)$$

If the eigenfunctions are also separately normalized, so that $\int \psi_m^* \psi_m \, d\tau = \int \psi_n^* \psi_n \, d\tau = 1$, then

$$\int \psi_n^* \psi_m \, d\tau = \delta_{nm} \qquad (2.17)$$

where $\delta_{nm} = 0$ when $n \neq m$ and 1 when $n = m$. Eigenfunctions that are both orthogonal and normalized are said to be orthonormal.

The first three postulates capture the essence of the operator form of quantum mechanics. There are further postulates, but these three are the most important. They firmly establish the existence of the wavefunction and its status as a complete description of the state of a quantum particle, the replacement of the values of observable quantities (classical mechanics) with their corresponding operators (quantum mechanics) and they provide a recipe for using the wavefunctions and operators to calculate the values of the observables. All the rest follows.

Complementary observables

We saw above that if the expectation value of an operator is calculated using one of its eigenfunctions, the result is equal to the corresponding eigenvalue. If the quantum system of interest is specified by the normalized wavefunction ψ_n, then the measurement of some property $\langle A_n \rangle$ (position, momentum, energy etc) requires the evaluation of the integral $\int \psi_n^* \hat{A} \psi_n \, d\tau$ which, as we have seen, is equal to a_n if ψ_n is an eigenfunction of $\hat{A}$. There is nothing inherent in this process to limit the precision with which $\langle A_n \rangle$ (which is equal to a_n) can be determined, i.e. there is in principle no uncertainty associated with $\langle A_n \rangle$.

Suppose we wish to determine a second property, say $\langle B_n \rangle$, of the state described by the wavefunction ψ_n. Clearly, from postulate 3, $\langle B_n \rangle = \int \psi_n^* \hat{B} \psi_n \, d\tau$ where $\hat{B}$ is the operator corresponding to this second observable (we have assumed that ψ_n is normalized). If we wish to determine $\langle B_n \rangle$ simultaneously with the same arbitrarily high precision as we determined $\langle A_n \rangle$, then we require $\langle B_n \rangle = b_n$ where b_n is the corresponding eigenvalue. Hence, ψ_n must be a simultaneous eigenfunction of $\hat{B}$.

We conclude that in order to measure simultaneously two different observables of a quantum state to arbitrary precision, the wavefunction describing that quantum state must be a simultaneous eigenfunction of both of the operators corresponding to the observables: ψ_n must be a simultaneous eigenfunction of both $\hat{A}$ and $\hat{B}$. What does this imply? Well, consider the action of the commutator $[\hat{A}, \hat{B}]$ on ψ_n:

$$
\begin{aligned}
[\hat{A}, \hat{B}]\psi_n &= (\hat{A}\hat{B} - \hat{B}\hat{A})\psi_n = \hat{A}\hat{B}\psi_n - \hat{B}\hat{A}\psi_n \\
&= \hat{A}b_n\psi_n - \hat{B}a_n\psi_n \\
&= b_n\hat{A}\psi_n - a_n\hat{B}\psi_n \\
&= b_n a_n \psi_n - a_n b_n \psi_n = 0,
\end{aligned}
\tag{2.18}
$$

i.e. $[\hat{A}, \hat{B}] = 0$ and the operators commute.

Quite simply, we can measure the values of two observables of some quantum state to arbitrary precision only if their corresponding operators commute. We have already seen that simultaneous determination of the position and momentum of a quantum particle to arbitrary precision is not possible (uncertainty principle) because their operators do not commute. Such observables are said to be complementary: we can measure one or the other with arbitrarily high precision but not both simultaneously. This is an extremely important point, to which we will be returning later.

2.3 STATE VECTORS IN HILBERT SPACE

The Dirac bracket notation

Integral equations like eqns (2.12)–(2.14) crop up all the time in quantum mechanics, and it quickly becomes tedious to keep writing them out in full. An extremely elegant shorthand notation was introduced by Paul Dirac in 1939 which not only serves to reduce the tedium but also provides considerable additional mathematical insight.

Instead of dealing with the wavefunction ψ_n, we define a related quantum 'state', denoted $|\psi_n\rangle$, of which ψ_n is just an alternative representation. The quantity $|\psi_n\rangle$ completely describes the state of a quantum system associated with a set of one or more quantum numbers denoted collectively by n (postulate 1). It has all the properties and all the significance that we have so far invested in ψ_n, including the latter's interpretation as a probability amplitude whose modulus squared gives the probability density of the quantum particle in a given region of space. $|\psi_n\rangle$ is called variously a 'ket', 'ket vector', 'state' or 'state vector', the last hinting at a significance that we will explore further in the next section. We will use the term state vector here.

The complex conjugate of $|\psi_n\rangle$ is the 'bra' $\langle\psi_n|$. When a 'bra' is combined with a 'ket', the result is a 'bracket'. The all-important integrals that quantum theory routinely requires us to deal with are represented as follows:

$$\int \psi_n^* \hat{A} \psi_m \, d\tau \equiv \langle\psi_n|\hat{A}|\psi_m\rangle$$

(2.19)

$$\int \psi_n^* \psi_m \, d\tau \equiv \langle\psi_n|\psi_m\rangle.$$

In terms of the state vector $|\psi_n\rangle$ the eigenvalue equation equivalent to $\hat{A}\psi_n = a_n\psi_n$ is

$$\hat{A}|\psi_n\rangle = a_n|\psi_n\rangle$$

(2.20)

and state vectors that satisfy such an equation are said to be eigenstates of the operator $\hat{A}$.

Hilbert space

If the wavefunctions and state vectors of quantum theory were functions only in 'ordinary' three-dimensional Euclidean space, then the theory would be nowhere near as abstract as it is. However, as we saw in Section 1.6, both experiment and the relativistic version of the quantum theory of the electron demand the existence of a spin 'coordinate'. This is a new, internal, coordinate of the electron which is not related in any way to the three conventional Cartesian coordinates. (If the electron were really a tiny particle spinning on its axis, it would be possible to describe this motion using Cartesian coordinates.)

If the three dimensions of Euclidean space are insufficient to describe quantum particles, what kind of space *is* required? The answer to this question is relatively simple: quantum particles are particles in Hilbert space, named after the mathematician David Hilbert (of whom John von Neumann was a student). Hilbert space is an abstract, mathematical space consisting, in principle, of an infinite number of dimensions. We take as many dimensions as are needed to specify completely the state of a quantum particle. Our use of this concept poses no real problems provided we keep ourselves to the necessary mathematical manipulations in Hilbert space. We can try to visualize what these manipulations mean in terms of 'everyday' Euclidean geometry, but we must always remember what it is that is being represented.

Obviously, there must be some kind of relationship between some of the dimensions in Hilbert space and those of Euclidean space, since we can write down many wavefunctions in ordinary Cartesian coordinates.

In fact, Euclidean space is a small sub-space of Hilbert space. We will see below that state vectors have all the properties that we tend to associate with vectors in classical physics, except that classical vectors are vectors in Euclidean space whereas state vectors are vectors in Hilbert space.

The expansion theorem

In our discussion of postulate 3, we indicated that when the wavefunction (or state vector) is not an eigenfunction (eigenstate) of an operator, the resulting expectation value of the operator gives only the 'mean value' of the observable. Let us find out what is meant by this by supposing that we wish to find the expectation value $\langle A \rangle$ of some operator $\hat{A}$ using some state vector $|\Psi\rangle$ which is not an eigenstate of $\hat{A}$. Clearly, we can proceed only if we can somehow recast the problem in terms of the eigenstates of $\hat{A}$, since we know their eigenvalues and we can make use of properties such as orthonormality which we know such eigenstates possess. Here we find it extremely helpful to make use of an important theorem of quantum mechanics, known as the expansion theorem or the superposition principle: an arbitrary, well behaved state vector can be expanded as a linear superposition of the complete set of eigenstates of any hermitian operator.

By 'well behaved' we mean that the state vector has properties closely related to those of the eigenstates, so that it has potentially the same kind of physical interpretation, and conforms to the same set of boundary conditions. By 'complete' we mean that the full set of eigenstates of the hermitian operator are needed to specify completely the state $|\Psi\rangle$. Such a full set is sometimes called a *basis set* and the individual eigenstates are referred to as basis states. We should note that although we have defined this theorem to be one of quantum mechanics, it is actually related to a quite general mathematical theorem which is used to expand an arbitrary function as a series of simpler functions. A good example is the Fourier series, in which a complicated function can be expressed as the sum of a set of simple sine or cosine functions. What makes this principle applicable in quantum mechanics is the wave nature of quantum particles.

To make life simple, we will assume that only two eigenstates, $|\psi_m\rangle$ and $|\psi_n\rangle$, are needed to specify completely the state vector $|\Psi\rangle$, i.e. we need a basis set of only two eigenstates. The expansion theorem suggests that we mix these two eigenstates together in some proportion that has yet to be determined, so we write

$$|\Psi\rangle = c_m|\psi_m\rangle + c_n|\psi_n\rangle \tag{2.21}$$

where c_m and c_n are mixing coefficients which indicate how much of each eigenstate is present in the mixture.

The expectation value of $\hat{A}$ in terms of $|\Psi\rangle$ is given by

$$\langle A \rangle = \frac{\langle \Psi | \hat{A} | \Psi \rangle}{\langle \Psi | \Psi \rangle}. \tag{2.22}$$

Let us evaluate this expression in stages. The effect of the operator $\hat{A}$ on $|\Psi\rangle$ is given by

$$\hat{A}|\Psi\rangle = c_m \hat{A}|\psi_m\rangle + c_n \hat{A}|\psi_n\rangle \tag{2.23}$$

where we have taken the coefficients to the left of the operator since they are just numbers. The complex conjugate of $|\Psi\rangle$ is the bra, $\langle\Psi|$, given by

$$\langle \Psi | = c_m^* \langle \psi_m | + c_n^* \langle \psi_n | \tag{2.24}$$

and so

$$\begin{aligned}
\langle \Psi | \hat{A} | \Psi \rangle &= (c_m^* \langle \psi_m | + c_n^* \langle \psi_n |)(c_m \hat{A}|\psi_m\rangle + c_n \hat{A}|\psi_n\rangle) \\
&= |c_m|^2 \langle \psi_m | \hat{A} | \psi_m \rangle + c_m^* c_n \langle \psi_m | \hat{A} | \psi_n \rangle + \qquad (2.25) \\
&\quad c_m c_n^* \langle \psi_n | \hat{A} | \psi_m \rangle + |c_n|^2 \langle \psi_n | \hat{A} | \psi_n \rangle
\end{aligned}$$

where $|c_m|^2 = c_m^* c_m$ and $|c_n|^2 = c_n^* c_n$. We now know from the previous section that if $|\psi_m\rangle$ and $|\psi_n\rangle$ are normalized eigenstates of $\hat{A}$, then $\langle \psi_m | \hat{A} | \psi_m \rangle = a_m$ and $\langle \psi_n | \hat{A} | \psi_n \rangle = a_n$. We can quickly deduce that $\langle \psi_m | \hat{A} | \psi_n \rangle = \langle \psi_m | a_n | \psi_n \rangle = a_n \langle \psi_m | \psi_n \rangle = 0$, since the eigenstates are also orthogonal. Simlarly, $\langle \psi_n | \hat{A} | \psi_m \rangle = 0$ and

$$\langle \Psi | \hat{A} | \Psi \rangle = |c_m|^2 a_m + |c_n|^2 a_n. \tag{2.26}$$

Much the same kind of procedure can be followed to show that $\langle \Psi | \Psi \rangle = |c_m|^2 + |c_n|^2 = 1$ if $|\Psi\rangle$ is normalized.

Equation (2.26) shows that the expectation value $\langle A \rangle$ is quite literally the 'mean value' of the observable: it is in fact a weighted average of the eigenvalues of the two eigenstates that make up $|\Psi\rangle$.

We can see what is going on here a little more clearly by drawing up the following:

$$\begin{bmatrix} \langle \psi_m | \hat{A} | \psi_m \rangle & \langle \psi_m | \hat{A} | \psi_n \rangle \\ \langle \psi_n | \hat{A} | \psi_m \rangle & \langle \psi_n | \hat{A} | \psi_n \rangle \end{bmatrix} = \begin{bmatrix} a_m & 0 \\ 0 & a_n \end{bmatrix}. \tag{2.27}$$

Each element within square brackets on the left has a corresponding element on the right. Equation (2.27) is, of course, a matrix equation.* The elements $\langle \psi_m | \hat{A} | \psi_m \rangle$ and $\langle \psi_n | \hat{A} | \psi_n \rangle$ are therefore sometimes referred to as *diagonal matrix elements* (because they lie along the diagonal from the top left to the bottom right of the matrix). The elements $\langle \psi_m | \hat{A} | \psi_n \rangle$ and $\langle \psi_n | \hat{A} | \psi_m \rangle$ are sometimes called *off-diagonal matrix elements*. Clearly, if the state vectors $| \psi_m \rangle$ and $| \psi_n \rangle$ are eigenstates of $\hat{A}$, the off-diagonal matrix elements are zero.

The expansion theorem is extremely important. Almost any problem in quantum mechanics for which the functional form of the state vector or wavefunction cannot easily be deduced can be solved in principle by expanding the state vector as a linear superposition of eigenstates of the operator corresponding to the property we are interested in (often the total energy).

We are completely free to choose whatever set of eigenstates we like, but it makes sense to choose ones that bear some resemblance to the problem we are trying to solve. For example, the wavefunctions of electrons in molecules can be modelled using a basis of atomic wavefunctions. Unfortunately, because we need *all* of the wavefunctions to form a complete set, including so-called continuum wavefunctions associated with ionized states, it is very difficult to reproduce the wavefunction of interest exactly. However, if we are happy to accept a small 'truncation' error, a judicious choice of basis will mean that we can get away with a much smaller number of basis states.

Projection amplitudes

What are the coefficients c_m and c_n in eqn (2.21)? We can answer this question in a quite straightforward manner by multiplying eqn (2.21) from the left by $\langle \psi_m |$:

$$\langle \psi_m | \Psi \rangle = c_m \langle \psi_m | \psi_m \rangle + c_n \langle \psi_m | \psi_n \rangle = c_m. \tag{2.28}$$

This follows because $\langle \psi_m | \psi_m \rangle = 1$ (normalization) and $\langle \psi_m | \psi_n \rangle = 0$ (orthogonality). Similarly,

$$\langle \psi_n | \Psi \rangle = c_n. \tag{2.29}$$

These expressions for c_m and c_n can be used in eqn (2.21) to give

$$| \Psi \rangle = | \psi_m \rangle \langle \psi_m | \Psi \rangle + | \psi_n \rangle \langle \psi_n | \Psi \rangle. \tag{2.30}$$

---

* Note that this should not be taken to imply that a ħ̶ These matrix elements are integrals derived from the operator (Schrödinger) form of quantum mechanics: the matrix elements of matrix mechanics are quite different (although they are related).

The terms $\langle \psi_m | \Psi \rangle$ and $\langle \psi_n | \Psi \rangle$ are sometimes called *inner products* or *projection amplitudes*.

The use of the term projection is very evocative. It cements the relationship between the ideas of vectors in classical physics and quantum state vectors. Imagine a vector **v** pointing in some arbitrary direction in Euclidean space. Such a vector might represent the instantaneous motion of a train; the train is going in a specific direction with a certain velocity. We draw an arrow to represent the direction of the vector and the length of the arrow represents its magnitude (Fig. 2.1). We define the vector to have a length of unity in some arbitrary unit system. Now suppose we want to 'map out' this vector in terms of its components along Cartesian axes (say x and y, as shown in the figure). We resolve the vector **v** into two orthogonal components; each component is also a vector which we represent as a coefficient multiplied by the unit vector corresponding to that particular direction. In other words,

$$\mathbf{v} = v_x \mathbf{i} + v_y \mathbf{j} \tag{2.31}$$

where v_x and v_y are the coefficients and **i** and **j** are the corresponding unit vectors in the x and y directions respectively.

The coefficients v_x and v_y are the projections of the vector **v** onto the x and y axes. They can be calculated as the inner products of **v** and the unit vectors:

$$v_x = (\mathbf{i} \cdot \mathbf{v}) = |\mathbf{i}| \, |\mathbf{v}| \cos\alpha = \cos\alpha$$
$$v_y = (\mathbf{j} \cdot \mathbf{v}) = |\mathbf{j}| \, |\mathbf{v}| \cos(90 - \alpha) = \sin\alpha \tag{2.32}$$

where α is the angle between the direction of **v** and the x axis. The

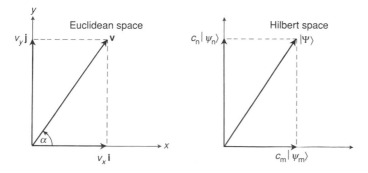

Fig. 2.1 Comparison of unit vectors in Euclidean space and state vectors in Hilbert space.

modulus $|\mathbf{v}|$ represents the magnitude of $\mathbf{v}$ and is independent of its direction. Obviously, $|\mathbf{v}| = |\mathbf{i}| = |\mathbf{j}| = 1$ since these are defined as unit vectors. Combining eqns (2.31) and (2.32) gives

$$\mathbf{v} = \mathbf{i}(\mathbf{i}\cdot\mathbf{v}) + \mathbf{j}(\mathbf{j}\cdot\mathbf{v}). \qquad (2.33)$$

Now compare eqns (2.30) and (2.33). In eqn (2.33), an arbitrary unit vector in Euclidean space is decomposed into two orthogonal components. The contribution that each orthogonal unit vector makes is determined by the magnitude of the projection of the arbitrary vector along the direction corresponding to that of the orthogonal unit vector. This projection is calculated from the inner product of the orthogonal unit vector and the arbitrary vector. In eqn (2.30), an arbitrary state vector in Hilbert space is decomposed into two orthogonal components. The contribution that each eigenstate makes is determined by the magnitude of the projection amplitude of the state vector along the direction corresponding to that eigenstate. This projection is calculated from the inner product or projection amplitude of the eigenstate and the arbitrary state vector. The analogy is complete: *state vectors are the unit vectors of Hilbert space.*

State vectors and classical unit vectors

Euclidean space is three-dimensional and so only three unit vectors, usually symbolized by $\mathbf{i}$, $\mathbf{j}$ and $\mathbf{k}$, are needed to specify completely an arbitrary vector. In contrast, Hilbert space has as many dimensions as we need to specify completely an arbitrary state vector. The parallels between state vectors and unit vectors can be clearly drawn out by considering their properties.

Firstly, they provide unique *representations* for quantum mechanical state vectors and classical vectors:

quantum $\qquad |\Psi\rangle = |\psi_m\rangle\langle\psi_m|\Psi\rangle + |\psi_n\rangle\langle\psi_n|\Psi\rangle,$

classical $\qquad \mathbf{v} = \mathbf{i}(\mathbf{i}\cdot\mathbf{v}) + \mathbf{j}(\mathbf{j}\cdot\mathbf{v}).$

In general, these representations can be written:

quantum $\qquad |\Psi\rangle = \sum_{s} |\psi_s\rangle\langle\psi_s|\Psi\rangle,$

classical $\qquad \mathbf{v} = \sum_{s=i,j,k} \mathbf{s}(\mathbf{s}\cdot\mathbf{v}).$

Secondly, both state vectors and unit vectors have the property of *orthogonality*:

quantum $\qquad \langle\psi_m|\psi_n\rangle = 0,$

classical $\qquad\qquad (\mathbf{i}\cdot\mathbf{j}) = \cos 90° = 0.$

Finally, both representations have the property of *completeness*:

quantum $\qquad\qquad |\langle \psi_{\mathrm{m}} | \Psi \rangle|^2 + |\langle \psi_{\mathrm{n}} | \Psi \rangle|^2 = 1,$

classical $\qquad\quad (\mathbf{i}\cdot\mathbf{v})^2 + (\mathbf{j}\cdot\mathbf{v})^2 = \cos^2\alpha + \sin^2\alpha = 1.$

The idea of the state vector thus brings with it many of the mathematical properties we associate with vectors in classical physics. To some extent, this is very helpful. Because we are familiar with the idea of classical vectors and can visualize what they are and how they combine, we are provided with an interpretation of state vectors that is intrinsically appealing. However, we should be under no illusions. The state vectors have properties that classical vectors can never have. The state vectors are vectors in a mathematically defined space and they can show interference effects. Whereas we can 'measure' vectors in classical physics, we cannot measure state vectors directly: only the modulus-squared of the state vector is accessible from experiment. The analogy between state vectors and unit vectors is a *mathematical* one: it offers us no help in deciding what a state vector *is*.

2.4 THE PAULI PRINCIPLE

The problem of explaining extra lines in the hydrogen atom spectrum was solved by introducing the idea of electron spin and its justification through the Dirac equation. All seemed to be well, but when physicists looked closely at the spectra of atoms containing more than one electron, they found that some lines seemed to be *missing*. There are two possible explanations for the non-appearance of an otherwise expected atomic line. Either the transition between quantum states responsible for the line is for some reason extremely weak or something is wrong with the theory and the quantum states themselves are not really there. The former explanation is often invoked in modern atomic and molecular spectroscopy: there is usually something about the state vectors involved that makes the transition a 'forbidden' one. However, it was the latter explanation that was used in the mid-1920s to explain the 'missing' lines in the spectrum of atomic helium.

The exclusion principle

The state of an electron in an atom is completely specified by the set of four quantum numbers n, l, m_1 and m_s. We know that as we add more and more electrons to an atom, they tend to occupy higher and higher energy orbitals. For example, once we have two electrons in the $1s$

($n = 1, l = 0, m_1 = 0$) orbital, that orbital is 'filled' and further electrons must go into the higher energy $2s$ or $2p$ orbitals. Why? There was nothing in the quantum theory of the early 1920s to suggest that two electrons could not possess the same values of the four quantum numbers. If there is no such restriction, why do the electrons not just all fall (or 'condense') into the lowest energy orbital?

In 1925, the Austrian physicist Wolfgang Pauli proposed that it was necessary to accept, as a general rule, that no two electrons could possess the same set of values of the four quantum numbers. Thus, as we feed electrons into an atom, the best we can do is get two electrons into any one orbital. For example, an electron going into a $1s$ orbital has $n = 1$, $l = 0$, $m_1 = 0$ and, for the sake of argument, we suppose it has $m_s = +\frac{1}{2}$ (spin-up). A second electron can go into the same orbital provided it adopts a spin-down, $m_s = -\frac{1}{2}$, orientation. An orbital can hold a maximum of two electrons with their spins *paired*. Further electrons must go into a higher energy orbital.

This is the Pauli exclusion principle, and its consequences are known to anyone who has studied elementary chemistry or physics. The exclusion principle helps to explain the periodic table of the elements. It provides the basis for understanding chemical bonding. In essence, it underpins all of chemistry. But knowing the principle, and using it to explain other aspects of the physical world brings us no closer to understanding why electrons have this property. We will see below that the exclusion principle is but one part of the more general Pauli principle, in turn based on the notion of indistinguishability.

Indistinguishable particles

If I were to acquire two apples that had exactly the same shape, size and coloration, and I were to place them side by side on my desk, we might, perhaps, agree that these apples are indistinguishable. But would this be strictly true? After all, I can use a metre rule to measure off the distances to each apple from the front and left-hand edges of my desk (the x and y axes) and note that apple 1 has coordinates x_1, y_1 and that apple 2 has coordinates x_2, y_2. These two sets of coordinates must be different, otherwise the apples would occupy the same space (they would be the same apple). Thus, the apples are distinguishable because they occupy different regions of space.

However, electrons are quantum wave–particles. We saw in Section 1.4 why we must abandon the idea that we can somehow keep track of an electron as it orbits the nucleus of an atom. Instead, we tend to think of electrons in terms of delocalized probability densities and the three-dimensional shapes of their density 'maps' correspond to our familiar pictures of atomic orbitals. If two electrons occupy an atomic orbital, how can we distinguish between them? We cannot now measure the coordinates of the two electrons in the same way that we can measure the coordinates of apples on my desk. The fact is that the electrons, like all quantum wave–particles, are indistinguishable.

The statistics of counting distinguishable particles are completely different from those for counting indistinguishable particles. Remember from Section 1.1 (and Appendix A) that Planck decided to use Boltzmann's statistical approach in deriving his radiation law. However, instead of assuming his energy elements to be distinguishable (as Boltzmann had always assumed when applying his methods to atoms and molecules) Planck purposefully made them indistinguishable. Paul Ehrenfest pointed out in 1911 that in doing this, Planck had given his quanta properties that were simply impossible for classical particles. Of course, photons and electrons are not classical particles. They possess wave-like properties too, and these properties lead to behaviour that is completely counter-intuitive if we try to think of photons and electrons as tiny, self-contained particles.

Before we tie ourselves in knots, let us take a look at what indistinguishability means in terms of state vectors. The state vector for a two-particle state (a state consisting of two electrons, for example) is just the product of the state vectors of the two particles. Thus, if particle 1 is described by the state vector $|\psi_m\rangle$, and particle 2 is described by the state vector $|\psi_n\rangle$, the appropriate product state can be written $|\psi_m^1\rangle|\psi_n^2\rangle$, where the superscripts indicate the individual particles.

But these particles are supposed to be indistinguishable. We have labelled the particles as 1 and 2 but if they are indeed indistinguishable we have no way of telling them apart. We can certainly distinguish between the possible quantum states $|\psi_m\rangle$ and $|\psi_n\rangle$ since they can correspond to states with different quantum numbers, energies, angular momenta etc, but we cannot tell experimentally which particle is in which state. Thus, the product $|\psi_n^1\rangle|\psi_m^2\rangle$ is just as acceptable as $|\psi_m^1\rangle|\psi_n^2\rangle$. (Note that the order in which we write the functions down is irrelevant, $|\psi_m^1\rangle|\psi_n^2\rangle = |\psi_n^2\rangle|\psi_m^1\rangle$.)

Because both of these product state vectors are equally 'correct', we have to assume that we can write a total two-particle state vector, denoted $|\Psi^{12}\rangle$, as a linear superposition of both these possibilities:

$$|\Psi^{12}\rangle = c_{mn}|\psi_m^1\rangle|\psi_n^2\rangle + c_{nm}|\psi_n^1\rangle|\psi_m^2\rangle. \qquad (2.34)$$

This mixture must contain equal proportions of each product state (because they must be equally possible), and so it follows that $|c_{mn}| = |c_{nm}|$. Furthermore, if we assume that $|\Psi^{12}\rangle$ is normalized, $|c_{mn}|^2 + |c_{nm}|^2 = 1$, and so $|c_{mn}| = |c_{nm}| = 1/\sqrt{2}$.

The only quantities accessible to us through experiment are the modulus-squares of the state vectors (the probability densities). This is why we have taken care to ensure that our conclusion about what the relationship between the coefficients must be is based on their moduli. However, the signs of these coefficients can certainly have important, measureable effects, as we will see below.

Fermions and bosons

There are obviously two ways of arriving at $|c_{mn}| = |c_{nm}|$. Either $c_{mn} = -c_{nm}$, or $c_{mn} = c_{nm}$. Let us take a look at the first possibility. If $c_{mn} = -c_{nm}$, and $|c_{mn}| = |c_{nm}| = 1/\sqrt{2}$, then the total two-particle state vector has the form:

$$|\Psi^{12}\rangle = \frac{1}{\sqrt{2}} \left(|\psi_m^1\rangle |\psi_n^2\rangle - |\psi_n^1\rangle |\psi_m^2\rangle \right). \qquad (2.35)$$

Now let us exchange the particles, so that what was labelled 1 becomes labelled as 2 and vice versa. We find that

$$\begin{aligned} |\Psi^{21}\rangle &= \frac{1}{\sqrt{2}} \left(|\psi_m^2\rangle |\psi_n^1\rangle - |\psi_n^2\rangle |\psi_m^1\rangle \right) \\ &= \frac{1}{\sqrt{2}} \left(|\psi_n^1\rangle |\psi_m^2\rangle - |\psi_m^1\rangle |\psi_n^2\rangle \right) \\ &= -|\Psi^{12}\rangle \end{aligned} \qquad (2.36)$$

The state vector $|\Psi^{12}\rangle$ is said to be antisymmetric (it changes sign) on the exchange of the particles. Note that, since only the sign of the state vector changes, its modulus-squared is indistinguishable from that obtained when the particles are exchanged. The particles are experimentally indistinguishable — their exchange should not (and does not) make any difference to quantities we can measure experimentally.

Now consider what happens if we try to put two quantum particles into the same state, i.e. we make $|\psi_n\rangle = |\psi_m\rangle$. In this situation,

$$|\Psi^{12}\rangle = \frac{1}{\sqrt{2}} \left(|\psi_m^1\rangle |\psi_m^2\rangle - |\psi_m^1\rangle |\psi_m^2\rangle \right) = 0. \qquad (2.37)$$

We conclude that quantum particles whose two-particle state vectors are antisymmetric to exchange are forbidden from occupying the same quantum state. This is just what was stated above for electrons, but our conclusion here applies to all particles with antisymmetric two-particle state vectors. Such particles are collectively called fermions, and have half-integral spin quantum numbers. Examples include electrons, protons, neutrons and some atomic nuclei.

The second possibility is $c_{mn} = c_{nm}$, or

$$| \Psi^{12} \rangle = \frac{1}{\sqrt{2}} \left(|\psi_m^1\rangle |\psi_n^2\rangle + |\psi_n^1\rangle |\psi_m^2\rangle \right). \tag{2.38}$$

In this case, the exchange of the two particles produces

$$\begin{aligned} | \Psi^{21} \rangle &= \frac{1}{\sqrt{2}} \left(|\psi_m^2\rangle |\psi_n^1\rangle + |\psi_n^2\rangle |\psi_m^1\rangle \right) \\ &= \frac{1}{\sqrt{2}} \left(|\psi_m^1\rangle |\psi_n^2\rangle + |\psi_n^1\rangle |\psi_m^2\rangle \right) \\ &= | \Psi^{12} \rangle. \end{aligned} \tag{2.39}$$

i.e. the state vector $| \Psi^{12} \rangle$ is symmetric (it does not change sign) on the exchange of particles. Particles whose two-particle state vectors possess this property are known as bosons, and have zero or integral spin quantum numbers. Examples include photons and some atomic nuclei.

The Pauli principle applies to all quantum particles. This principle states that particles with half-integral spin quantum numbers — fermions — must have two-particle (or, in general, many-particle) state vectors that are antisymmetric with respect to the pairwise interchange of particles. Particles with integral spin quantum numbers — bosons — must have symmetric many-particle state vectors. The Pauli exclusion principle is an extension of the Pauli principle as applied to electrons: the requirement for an antisymmetric state vector for electrons means that electrons are excluded from occupying the same quantum state. These symmetry requirements arise naturally when the effects of special relativity are introduced in the quantum mechanics of many-particle quantum states, as was shown by Pauli himself.

Readers might be forgiven for thinking that the Pauli principle provides yet another layer of mysterious formalism for quantum systems containing many particles on top of an already quite impenetrable formalism for single particles. I sympathize, but there are, in fact, not that many mysteries. At the heart of the Pauli principle lies the indistinguishability of all quantum particles, with fermions differing from bosons in the symmetry properties of their many-particle state vectors. As we have seen, the assumption of indistinguishable energy elements was a necessary part of Planck's 'act of desperation' which led ultimately to the development of quantum theory.

Indistinguishability is a property of quantum particles that is intrinsically linked to their wave-particle nature, as is the position–momentum commutation relation and Heisenberg's uncertainty principle. All these problems are one problem.

2.5 THE POLARIZATION PROPERTIES OF PHOTONS

Readers might be justifiably anxious that, in a chapter entitled 'Putting it into practice', we seem to have devoted ourselves to the mathematical formalism of quantum theory and have paid scant attention to its application to 'real' systems. However, now that we have enough elements of the formalism in place, we can begin to look at how it should be used. Remember that in this chapter, we are accepting without question the postulates on which the orthodox form of the theory is based: detailed discussion of the validity of these postulates is deferred until later chapters.

This section actually serves three purposes. Firstly, it provides us with an opportunity to become more familiar with the way in which state vectors are manipulated and related to measurable quantities. Secondly, by focusing on the polarization properties of photons, we are establishing the basic background needed to interpret the important experiments which are described in Chapter 4. Finally, since simple experiments with polarized light are relatively easy to perform (and imagine), we can use them to highlight some curious quantum phenomena in a fairly straightforward manner.

Linear polarization

We will begin with a discussion of the polarization properties of light that is based almost entirely on classical concepts. In the seventeenth century, Isaac Newton noticed that there appear to be two different 'types' of light, but it was the Dutch scientist Christian Huygens and, later, Thomas Young who produced an explanation. In terms of Maxwell's theory of electromagnetism, the electric vectors of transverse light waves confined to oscillating in one dimension only (plane waves) can take up two possible orientations that are mutually orthogonal and also perpendicular to the direction of propagation. We will assume that the direction of propagation of some plane light wave is along the z-axis. Thus, in vertically polarized light, the electric vector is confined to oscillate only in the x direction and in horizontally polarized light the electric vector is confined to oscillate in the y direction (see Fig. 2.2).

Most people are familiar with polarizing filters — pieces of plastic film (often called 'Polaroid' film after the manufacturer's trademark). This film consists of an array of polymer molecules which shows a preference for absorption of light along one specific axis. Imagine that we take two pieces of Polaroid film placed one on top of the other and we arrange for them to be illuminated from behind by a suitable light source. We arrange them so that the maximum amount of light is trans-

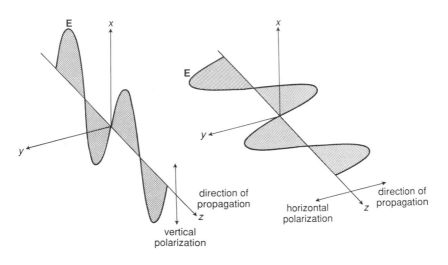

Fig. 2.2 Vertically and horizontally polarized plane electromagnetic waves. Only the electric vector of the waves is shown: the magnetic vector oscillates at right angles to both the electric vector and the direction of propagation.

mitted through both filters. Now we slowly rotate one of the filters through 90° and note how the intensity of transmitted light falls until, when the axes of maximum transmission of the two filters are at right angles, no light is transmitted at all (the filters are said to be 'crossed').

The eye is a powerful but non-quantitative light detector. If we were to measure the amount of light transmitted through both filters using a device such as a photomultiplier or a photodiode, we would discover that the intensity falls off according to the cosine-squared of the angle between the transmission axes of the filters. In other words,

$$I = I_0\cos^2\varphi \qquad (2.40)$$

where I is the transmitted intensity, I_0 is the intensity of light transmitted through the first filter and φ is the angle between the polarizers as defined in Fig. 2.3. This is Malus's law.

We can readily interpret this law using classical concepts. We can suppose that the first filter transmits light polarized predominantly along its axis of maximum transmission — the filter provides a source of (in this case we assume) vertically polarized light. As the angle between the two filters is changed, the second filter transmits only the component of the electric vector of the vertically polarized light which lies along its axis of

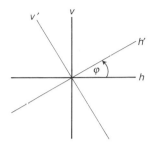

Fig. 2.3   Axis convention used in the analysis of linear polarization.

maximum transmission. This component is the projection of the electric vector of the vertically (v) polarized light onto the new vertical (v') axis, and therefore depends on the cosine of the angle between them. Since the intensity of light is proportional to the square of the modulus of the electric vector, the intensity of light transmitted through both filters varies as $\cos^2 \varphi$.

Polarization states

How should Malus's law be interpreted in terms of photons? According to quantum theory, we can assign each individual photon transmitted through the first filter to a state of vertical polarization. We denote such a state by the state vector $|\psi_v\rangle$. As the second filter is rotated, each photon is *projected* into a new state, $|\psi_v'\rangle$, with a *probability* equal to $\cos^2 \varphi$. The intensity of light transmitted through both filters depends on the number of photons detected. This number is determined by the probability that each photon is projected into the state $|\psi_v'\rangle$ and therefore transmitted by the second filter. We cannot predict if any one individual photon will be transmitted: we only know the probability with which it might be transmitted.

The projection probability is the modulus-squared of the corresponding projection amplitude:

$$|\langle \psi_v' | \psi_v \rangle|^2 = \cos^2 \varphi. \tag{2.41}$$

(It is a convention to write the final state, in this case $|\psi_v'\rangle$, on the left of the bracket and the initial state on the right.) Although we can use this relationship to tell us the absolute value of the projection amplitude ($|\cos\varphi|$) , we have no information at present on its sign.

We have assumed that the first filter produces vertically polarized light. However, we can use the axis convention given in Fig. 2.3 to deduce that

$$|\langle \psi_h' | \psi_h \rangle|^2 = \cos^2 \varphi$$

$$|\langle \psi_v' | \psi_h \rangle|^2 = \sin^2 \varphi \qquad (2.42)$$

$$|\langle \psi_h' | \psi_v \rangle|^2 = \sin^2 \varphi.$$

Exchanging the symbols in the brackets on the left-hand sides of eqns (2.42) implies the reverse processes, for which the results are identical. The properties of the quantum states themselves mean that $|\langle \psi_v | \psi_v \rangle|^2 = |\langle \psi_h | \psi_h \rangle|^2 = |\langle \psi_v' | \psi_v' \rangle|^2 = |\langle \psi_h' | \psi_h' \rangle|^2 = 1$ and $|\langle \psi_h | \psi_v \rangle|^2 = |\langle \psi_h' | \psi_v' \rangle|^2 = 0$. This follows from the assumption that the photon polarization states are eigenstates of some operator (they are properties that we can certainly observe) and are therefore orthonormal. They can also be deduced from five minutes' toying with the polarization filters.

Photon spin and circular polarization

Photons are bosons. They are quantum particles with spin quantum numbers $s = 1$. Like the electron, a photon can have only one value of s, and there are different ways that the photon spin can be 'aligned', corresponding to the different values of the magnetic spin quantum number m_s. For electrons, $s = \frac{1}{2}$ and so we have two possibilities: $m_s = +\frac{1}{2}$ (spin up) and $m_s = -\frac{1}{2}$ (spin down). As a general rule, quantum theory predicts the existence of states with values of m_s in the series s, $(s - 1)$, $(s - 2)$, . . ., 0, . . ., $-(s - 2)$, $-(s - 1)$, $-s$. This rule would lead us to predict that photons should have three possibilities for m_s, corresponding to $m_s = +1, 0$ and -1. However, relativistic quantum theory forbids an $m_s = 0$ component for particles travelling at the speed of light. This leaves us with just two possibilities and, by convention, we associate the $m_s = +1$ component with left circularly polarized light and the $m_s = -1$ component with right circularly polarized light. This makes sense if we remember that the spin property of a quantum particle is manifested as an intrinsic angular momentum. We define left circular as a counterclockwise rotation of the electric vector of light viewed along the direction of propagation and travelling towards the observer (see Fig. 2.4).

Although the spin property of a quantum particle should never be interpreted as if the particle were literally spinning on its axis, it is nevertheless manifested as an intrinsic angular momentum. Thus, a beam containing a large number of circularly polarized photons (such as in a laser beam) will impart a measurable torque to a target. However, this angular momentum is not a collective phenomenon: in the absorption of an

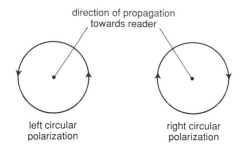

Fig. 2.4   Convention for circular polarization.

individual photon resulting in electron excitation in an atom or molecule, the angular momentum intrinsic to the photon is transferred to the excited electron and total angular momentum is conserved. That transfer has important, measurable effects on the spectrum of the absorbing species.

Linearly polarized light can be 'synthesized' from an equal mixture of left and right circularly polarized light. The evidence for this can be obtained by measuring the intensity of light transmitted through a polarizing filter. If we define the polarization states corresponding to left and right circular polarization as $|\psi_L\rangle$ and $|\psi_R\rangle$ respectively, we can do experiments to show that

$$|\langle \psi_v | \psi_L \rangle|^2 = \frac{1}{2}$$

$$|\langle \psi_h | \psi_L \rangle|^2 = \frac{1}{2}$$

(2.43)

$$|\langle \psi_v | \psi_R \rangle|^2 = \frac{1}{2}$$

$$|\langle \psi_h | \psi_R \rangle|^2 = \frac{1}{2}.$$

Thus, for example, left circularly polarized light incident on a polarizing filter will produce vertically polarized light with half the original intensity. In terms of photons, each left circularly polarized photon has a probability of $\frac{1}{2}$ of being projected into a state of vertical polarization (and hence being transmitted) and a probability of $\frac{1}{2}$ of being projected into a state of horizontal polarization (and hence being absorbed by the filter). We summarize these projection probabilities in Table 2.1.

Table 2.1   Projection probabilities, $|\langle \psi_f | \psi_i \rangle|^2$, for photon polarization states

Final state $|\psi_f\rangle$	Initial state $|\psi_i\rangle$					
	$|\psi_v\rangle$	$|\psi_h\rangle$	$|\psi_v'\rangle$	$|\psi_h'\rangle$	$|\psi_L\rangle$	$|\psi_R\rangle$
$|\psi_v\rangle$	1	0	$\cos^2\varphi$	$\sin^2\varphi$	1/2	1/2
$|\psi_h\rangle$	0	1	$\sin^2\varphi$	$\cos^2\varphi$	1/2	1/2
$|\psi_v'\rangle$	$\cos^2\varphi$	$\sin^2\varphi$	1	0	1/2	1/2
$|\psi_h'\rangle$	$\sin^2\varphi$	$\cos^2\varphi$	0	1	1/2	1/2
$|\psi_L\rangle$	1/2	1/2	1/2	1/2	1	0
$|\psi_R\rangle$	1/2	1/2	1/2	1/2	0	1

Basis states and projection amplitudes

It should by now have become obvious that these three sets of quantum states, $|\psi_v\rangle/|\psi_h\rangle$, $|\psi_v'\rangle/|\psi_h'\rangle$ and $|\psi_L\rangle/|\psi_R\rangle$, are all suitable representations for photon polarization states and they can therefore also be used as basis states. Although we use a convention to assign photons with $m_s = \pm 1$ to states of circular polarization, those states can, in turn, be expressed as linear superpositions of states of linear polarization.

For example, we can use the expansion theorem to write

$$|\psi_L\rangle = |\psi_v\rangle\langle\psi_v|\psi_L\rangle + |\psi_h\rangle\langle\psi_h|\psi_L\rangle \qquad (2.44)$$

and

$$|\psi_R\rangle = |\psi_v\rangle\langle\psi_v|\psi_R\rangle + |\psi_h\rangle\langle\psi_h|\psi_R\rangle. \qquad (2.45)$$

Similar expressions can be written for $|\psi_v\rangle$ and $|\psi_h\rangle$ in terms of $|\psi_L\rangle$ and $|\psi_R\rangle$. We can obviously go no further until we find expressions for the various projection amplitudes.

We can deduce the projection amplitudes for linear polarization states using the axis convention defined in Fig. 2.3 combined with a little vector algebra. From the analogy between state vectors and classical unit vectors described in Section 2.3 we note from Fig. 2.3 that

$$|\psi_v\rangle = |\psi_v'\rangle\cos\varphi + |\psi_h'\rangle\sin\varphi \qquad (2.46)$$

where $|\psi_v'\rangle$ and $|\psi_h'\rangle$ are the equivalent of unit vectors along the v', h' axes. Multiplying both sides of eqn (2.46) by $\langle\psi_v'|$ gives

$$\langle\psi_v'|\psi_v\rangle = \cos\varphi \qquad (2.47)$$

since $\langle\psi_v'|\psi_v'\rangle = 1$ and $\langle\psi_v'|\psi_h'\rangle = 0$. We can similarly deduce that

$$\langle \psi_h'|\psi_v\rangle - \sin\varphi$$

$$\langle \psi_v'|\psi_h\rangle = \cos(90 + \varphi) = -\sin\varphi \qquad (2.48)$$

$$\langle \psi_h'|\psi_h\rangle = \cos\varphi.$$

Obviously, it follows that $\langle \psi_v|\psi_v'\rangle = \langle \psi_v'|\psi_v\rangle$ etc., i.e. the projection amplitudes for linear polarization states are symmetric to the exchange of initial and final states. A quick glance at Table 2.1 reveals that these projection amplitudes are consistent with the corresponding projection probabilities, and hence they are consistent with experiment.

We now need to go on to consider projection amplitudes involving states of circular polarization. We said above that the $|\psi_v\rangle/|\psi_h\rangle$, $|\psi_v'\rangle/|\psi_h'\rangle$ and $|\psi_L\rangle/|\psi_R\rangle$ states serve as interchangeable sets of basis states for photon polarization. We can therefore express the state $|\psi_v'\rangle$ as a linear superposition of $|\psi_L\rangle$ and $|\psi_R\rangle$:

$$|\psi_v'\rangle = |\psi_L\rangle\langle\psi_L|\psi_v'\rangle + |\psi_R\rangle\langle\psi_R|\psi_v'\rangle. \qquad (2.49)$$

Multiplying both sides of this expression by $\langle\psi_v|$ gives

$$\langle\psi_v|\psi_v'\rangle = \langle\psi_v|\psi_L\rangle\langle\psi_L|\psi_v'\rangle + \langle\psi_v|\psi_R\rangle\langle\psi_R|\psi_v'\rangle = \cos\varphi. \qquad (2.50)$$

From Table 2.1, we know that $|\langle\psi_v|\psi_L\rangle| = |\langle\psi_L|\psi_v'\rangle| = |\langle\psi_v|\psi_R\rangle| = |\langle\psi_R|\psi_v'\rangle| = 1/\sqrt{2}$. There can be no way of reconciling the moduli of these projection amplitudes with eqn (2.50) without recognizing that some of the projection amplitudes must themselves be complex. We recall that

$$\cos\varphi = \frac{1}{2}(e^{i\varphi} + e^{-i\varphi})$$

$$= \frac{1}{2}e^{i\varphi} + \frac{1}{2}e^{-i\varphi} \qquad (2.51)$$

and so we (quite arbitrarily) identify the first term on the right-hand side of eqn (2.51) with the term $\langle\psi_v|\psi_L\rangle\langle\psi_L|\psi_v'\rangle$ and the second term with $\langle\psi_v|\psi_R\rangle\langle\psi_R|\psi_v'\rangle$, i.e.

$$\langle\psi_v|\psi_L\rangle\langle\psi_L|\psi_v'\rangle = \frac{1}{2}e^{i\varphi}$$

$$\qquad (2.52)$$

$$\langle\psi_v|\psi_R\rangle\langle\psi_R|\psi_v'\rangle = \frac{1}{2}e^{-i\varphi}.$$

Furthermore, since it is logical to associate the terms in $e^{\pm i\varphi}$ with those projection amplitudes involving the state $|\psi_v'\rangle$, we can decompose the expressions in eqns (2.52) to give the individual amplitudes as follows:

$$\langle \psi_v | \psi_L \rangle = \frac{1}{\sqrt{2}}, \quad \langle \psi_L | \psi_v' \rangle = \frac{1}{\sqrt{2}}\, e^{i\varphi}$$

$$\langle \psi_v | \psi_R \rangle = \frac{1}{\sqrt{2}}, \quad \langle \psi_R | \psi_v' \rangle = \frac{1}{\sqrt{2}}\, e^{-i\varphi}.$$

(2.53)

Notice that $|\langle \psi_L | \psi_v' \rangle|^2 = |\langle \psi_R | \psi_v' \rangle|^2 = \frac{1}{2}$, as required. If this seems to be a completely arbitrary procedure (we could just as well have taken $\langle \psi_v | \psi_L \rangle \langle \psi_L | \psi_v' \rangle = e^{-i\varphi}/\sqrt{2}$), it is because this is exactly what it is. Remember that we have no way of knowing the 'actual' signs of the phase factors because that information is not revealed in experiments. However, we can adopt a *phase convention* which, if we stick to it rigorously, will always give results that are both internally consistent and consistent with experiment.

Using the phase convention determined by the choices made in eqns (2.52) and (2.53), we can use the same general procedure to deduce all the projection amplitudes for all of the basis states. They are collected in Table 2.2. Note from this table that our phase convention leads to $\langle \psi_f | \psi_i \rangle = \langle \psi_i | \psi_f \rangle^*$, where i and f are any of v, h, v', h', L and R. We will now make use of these projection amplitudes in our dicussion of quantum measurement.

2.6 QUANTUM MEASUREMENT

Polarizing filters like the ones used in the above discussion are actually not very efficient. Such a filter might transmit as little as 70 per cent of linearly polarized light through its axis of maximum transmission. This is an annoying problem, but we can overcome it by switching to an alternative kind of polarization analyser. One such alternative is a piece of calcite, a naturally occurring crystalline form of calcium carbonate.

Table 2.2   Projection amplitudes, $\langle \psi_f | \psi_i \rangle$, for photon polarization states

Final state $|\psi_f\rangle$	Initial state $|\psi_i\rangle$					
	$|\psi_v\rangle$	$|\psi_h\rangle$	$|\psi_v'\rangle$	$|\psi_h'\rangle$	$|\psi_L\rangle$	$|\psi_R\rangle$
$|\psi_v\rangle$	1	0	$\cos\varphi$	$\sin\varphi$	$1/\sqrt{2}$	$1/\sqrt{2}$
$|\psi_h\rangle$	0	1	$-\sin\varphi$	$\cos\varphi$	$i/\sqrt{2}$	$-i/\sqrt{2}$
$|\psi_v'\rangle$	$\cos\varphi$	$-\sin\varphi$	1	0	$e^{-i\varphi}/\sqrt{2}$	$e^{i\varphi}/\sqrt{2}$
$|\psi_h'\rangle$	$\sin\varphi$	$\cos\varphi$	0	1	$ie^{-i\varphi}/\sqrt{2}$	$-ie^{i\varphi}/\sqrt{2}$
$|\psi_L\rangle$	$1/\sqrt{2}$	$-i/\sqrt{2}$	$e^{i\varphi}/\sqrt{2}$	$-ie^{i\varphi}/\sqrt{2}$	1	0
$|\psi_R\rangle$	$1/\sqrt{2}$	$i/\sqrt{2}$	$e^{-i\varphi}/\sqrt{2}$	$ie^{-i\varphi}/\sqrt{2}$	0	1

Calcite is naturally birefringent; it has a crystal structure which has different refractive indices along two distinct planes. One offers an axis of maximum transmission for vertically polarized light and the other offers an axis of maximum transmission for horizontally polarized light. The vertical and horizontal components of light which is a mixture of polarizations are therefore physically separated by passage through the crystal, and their intensities can be measured separately. With careful machining, a calcite crystal can transmit virtually all of the light incident on it.

There are a number of ways of obtaining a source of left circularly polarized light. These vary from a standard (i.e. unpolarized) light source passed through an optical device known as a quarter-wave plate to an atomic source that relies on the quantum mechanics of photon emission. An example of the latter is a beam of atoms that are excited to some electronically excited state from which emission occurs. If angular momentum is to be conserved in the process, the emitted photon must carry away any excess angular momentum lost by the excited electron as it returns to a more stable quantum state. An appropriate choice of states between which the transition occurs can give rise to the emission of photons exclusively with $m_s = +1$. We will meet this kind of source again in Chapter 4.

A beam of left circularly polarized light entering a calcite crystal will split into two beams, one of vertical and one of horizontal polarization (see Fig. 2.5). We can use detectors (such as photomultipliers), placed in the paths of the emergent beams, to confirm that each has half the intensity of the initial beam. This is consistent with the photon projection amplitudes and probabilities we deduced in the last section.

But let us now reduce the intensity of the incident left circularly polarized light so that, on average, only one photon passes through the crystal at a time. What happens to the photon? It cannot split into two, one half following one path and the other half following the other path, because the photon is a 'fundamental' particle. Besides, if we really could

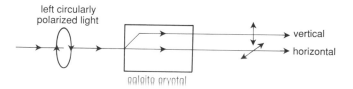

Fig. 2.5 A calcite crystal splits left circularly polarized light into two equal vertical and horizontal components.

split a photon in half, we would necessarily halve the energy (and, from $\varepsilon = h\nu$, the frequency). A simple experiment to measure the frequency of each transmitted photon confirms that it has the same value as the incident photon. If the photon follows only one path through the crystal, it must emerge *either* from the vertical 'channel' *or* from the horizontal 'channel'.

Measurement operators

We will see in the next chapter that the orthodox interpretation of quantum theory insists that the nature of the measuring apparatus, and the way it is set up, is of primary importance in the analysis of quantum systems. In our example developed above, we have put together a device to decompose the incident light into vertical and horizontal polarization components, which are then detected. The whole process of passing a photon through the calcite crystal and detecting it can be represented as an operator: the apparatus is, after all, a set of instructions to do various things to the state vector of the incident photon. We denote such a measurement operator as $\hat{M}$.

Now if we pass a vertically polarized photon through an ideal calcite crystal, it will emerge exclusively from the vertical channel and be detected. Thus, the effect of $\hat{M}$ operating on $|\psi_v\rangle$ is to produce the 'result', $|\psi_v\rangle$. Imagine we have the apparatus rigged so that a red light comes on if a photon is detected through the vertical channel. We may conscientiously enter this result in our laboratory notebook, perhaps representing it by writing R_v.

The above reasoning allows us to write

$$\hat{M}|\psi_v\rangle = R_v|\psi_v\rangle, \tag{2.54}$$

i.e. the state vector for vertical polarization is an eigenstate of the measurement operator, with eigenvalue R_v. This is merely a statement that the apparatus is set up to measure vertical polarization.

If we set up the apparatus so that a blue light comes on if a photon is detected through the horizontal channel, then we can use similar arguments to show that

$$\hat{M}|\psi_h\rangle = R_h|\psi_h\rangle \tag{2.55}$$

where R_h is the corresponding eigenvalue. Note that it is unnecessary for us to figure out the exact mathematical form of $\hat{M}$: its properties and effects on the state vectors are defined by the way we have the apparatus set up.

The state vector of a left circularly polarized photon can be expressed as a linear superposition of $|\psi_v\rangle$ and $|\psi_h\rangle$ which, using eqn (2.44) and the information given in Table 2.2, we can write as

$$|\psi_L\rangle = \frac{1}{\sqrt{2}}(|\psi_v\rangle + i|\psi_h\rangle).\tag{2.56}$$

The effect of passing a left circularly polarized photon through our measuring apparatus is therefore given by

$$\hat{M}|\psi_L\rangle = \frac{1}{\sqrt{2}}(\hat{M}|\psi_v\rangle + i\hat{M}|\psi_h\rangle) = \frac{1}{\sqrt{2}}(R_v|\psi_v\rangle + iR_h|\psi_h\rangle).\tag{2.57}$$

If we assume $|\psi_L\rangle$ to be normalized, the expectation value of the operator $\hat{M}$ for the state $|\psi_L\rangle$ is given by

$$\langle M_L\rangle = \langle\psi_L|\hat{M}|\psi_L\rangle = \frac{1}{\sqrt{2}}(\langle\psi_v| - i\langle\psi_h|)\frac{1}{\sqrt{2}}(R_v|\psi_v\rangle + iR_h|\psi_h\rangle)$$

$$= \frac{1}{2}(R_v\langle\psi_v|\psi_v\rangle + iR_h\langle\psi_v|\psi_h\rangle - iR_v\langle\psi_h|\psi_v\rangle + R_h\langle\psi_h|\psi_h\rangle)$$

$$= \frac{1}{2}(R_v + R_h).\tag{2.58}$$

This last equation indicates that we expect to see the red light and the blue light come on with equal probability. This does not mean that both lights are 'half on'. It means that, on average, the red light comes on for half of the photons detected and the blue light comes on for the other half. The theory does not allow us to predict with certainty which light will come on for a given incident photon.

Another way of looking at the measurement process is to say that, in order to obtain the result corresponding to the eigenvalue R_v, the initial photon state $|\psi_L\rangle$ must be projected into the state $|\psi_v\rangle$. The effect of $\hat{M}$ on the state vector is to yield the eigenvalue R_v. The corresponding projection probability, $|\langle\psi_v|\psi_L\rangle|^2$, is equal to $\frac{1}{2}$.

The 'collapse' of the wavefunction

An individual left circularly polarized photon must be detected to emerge from either the vertical or the horizontal channel of the calcite crystal. Prior to measurement, the quantum state of the photon can be described as a linear superposition of the two measurement eigenstates, eqn (2.56). After measurement, the photon is inferred to have been in one, and only one, of the measurement eigenstates. Somewhere along the way the state vector has changed from one consisting of two measurement *possibilities*

($|\psi_v\rangle$ *and* $|\psi_h\rangle$) to one *actuality* ($|\psi_v\rangle$ *or* $|\psi_h\rangle$). This process is known as the 'collapse', or 'reduction', of the wavefunction.

Readers might be inclined to think that we are labouring this point, but it goes directly to the heart of the meaning of quantum theory. What does eqn (2.56) actually represent? Might it not merely reflect the fact that left circularly polarized light is really a 50:50 mixture of vertically and horizontally polarized photons, and we use it because, prior to measurement, we are ignorant of the actual polarization state of any one individual photon? If this is the case, each individual photon is present in a pre-determined $|\psi_v\rangle$ or $|\psi_h\rangle$ state: each follows a predetermined path through the calcite crystal according to that state and is detected. Under such circumstances, the collapse of the wavefunction represents a sharpening of our knowledge of the state of the photon. Prior to measurement, the photon is in either $|\psi_v\rangle$ or $|\psi_h\rangle$, and the measurement merely tells us which.

Or does eqn (2.56) really reflect the fact that the linear polarization state of the photon is completely *undetermined* prior to measurement? In this case, the collapse of the wavefunction represents more than just a change in our state of knowledge of the system. In fact, this way of thinking requires a fundamental revision of our conception of the process of measurement compared with classical mechanics. For example, I assume the length of my desk to be a predetermined quantity. Although I accept that I have no knowledge of this quantity until I measure it, I do not assume that the very act of looking at my desk to locate its edges in space changes its length from an undetermined into a determined quantity. In classical physics, to have no knowledge of a physical quantity does not imply that it is not determined before a measurement is made.

We will see in the next chapter that Niels Bohr and his colleagues in Copenhagen favoured the interpretation that quantum measurement involves the projection of a previously undetermined quantum state into some measurement eigenstate. Albert Einstein and his colleagues stood firm for a completely deterministic approach. Their arguments led to the development of an important test case, which we will review in Chapter 4.

The time evolution operator

If the projection of some initial state vector into a measurement eigenstate is an intrinsic part of the measurement process, how is this projection described by the equations of quantum theory? The simple fact is that this process is not described at all. Since the act of measurement occurs within a finite time interval—a detector changes in time from some initial state to some final state – the place to look is the time-dependent Schrödinger equation, which we will now give:

$$i\hbar \frac{\partial}{\partial t} | \Psi \rangle = \hat{H} | \Psi \rangle. \qquad (2.59)$$

Like the time-independent Schrödinger equation, eqns (2.10) and (2.11), the time-dependent equation cannot be 'derived' in any rigorous way. In fact, it is often assumed as one of the postulates of quantum mechanics.

Integration of eqn (2.59) gives:

$$| \Psi \rangle = e^{-i\hat{H}t/\hbar} | \Psi_0 \rangle \qquad (2.60)$$

where $| \Psi_0 \rangle$ is the state vector at some initial time $t = 0$ and $| \Psi \rangle$ represents the state vector at some later time. (This can be readily checked by differentiating eqn (2.60) with respect to t.) If, at first sight, the exponential term in eqn (2.60) looks very strange, remember that we can expand an exponential as a power series:

$$e^{-i\hat{H}t/\hbar} = 1 - \frac{i\hat{H}t}{\hbar} - \frac{\hat{H}^2 t^2}{2\hbar^2} + \frac{i\hat{H}^3 t^3}{6\hbar^3} + \cdots \qquad (2.61)$$

from which it is more obvious how the terms in powers of $\hat{H}$ will operate on the state vector $| \Psi_0 \rangle$. The exponential term is called the *time-evolution operator* and is usually given the symbol $\hat{U}$. Equation (2.59) can therefore be written succinctly as:

$$| \Psi \rangle = \hat{U} | \Psi_0 \rangle, \qquad \hat{U} = e^{-i\hat{H}t/\hbar}. \qquad (2.62)$$

The most important lesson to be learned from eqn (2.62) is that *the time evolution of a quantum system is continuous and deterministic*. Once in the state $| \Psi_0 \rangle$, the quantum system will evolve continuously in time according to eqn (2.62). This equation cannot describe the discontinuous, indeterministic projection of $| \Psi \rangle$ into some measurement eigenstate.

As we described in Section 1.4, Max Born found that to describe transitions between quantum states, he had to *combine* the continuous, deterministic equations of Schrödinger's wave mechanics with the discontinuous, indeterministic quantum jumps. Similarly, in his theory of quantum measurement, John von Neumann combined the continuous, deterministic equations describing the time evolution of a quantum system with a discontinuous, indeterministic collapse of the wavefunction. The latter cannot be obtained from the former. We will look at the further implications of this approach in the next chapter, and we will be returning to von Neumann's theory of measurement in Chapter 5.

Some fun with photons

Before we leave this chapter, it is worth taking a quick look at some of the curious observations that can be made with photons, observations that we must attempt to interpret in terms of quantum theory. These will serve to whet the appetite for the fun which is to follow in the remaining chapters.

Thomas Young explained his observation of double-slit interference using a wave theory of light. A light beam of moderately high intensity incident on two closely spaced, narrow apertures produces an interference pattern consisting of bright and dark fringes. Now imagine that we reduce the light intensity of the source so that only one photon passes through the double-slit apparatus at a time, to impinge on some photographic film. Such experiments can, and have, been performed in the laboratory. After a significant number of photons have passed through, we find that the interference pattern is clearly visible (the equivalent experiment with electrons was described in Chapter 1, see Fig. 1.3).

If we assume that an individual photon must pass through one – and only one – slit, we should be able to repeat the experiment using a detector to discover which one. However, when such an experiment is done, we find that the interference pattern is replaced with a completely different pattern corresponding to the diffraction of light through the remaining open slit. The act of removing the detector and unblocking the second slit restores the interference pattern. We conclude that if a photon does pass through one slit, it must be somehow affected by the second, even though it cannot 'know' in advance that the second slit is open.

We have seen that a calcite crystal can be used to decompose left circularly polarized light into vertical and horizontal components. If we take an identical crystal, and orient it in the opposite sense, we can use it to recombine the vertical and horizontal components and reconstitute the left circularly polarized light (see Fig. 2.6). That such a reconstitution can be achieved has been proved in careful laboratory experiments.

Now suppose that an individual left circularly polarized photon passes through the first crystal and emerges from the vertical channel. The photon enters the vertical channel of the second crystal. At first glance there seems to be no way of obtaining a left circularly polarized photon out of this, and yet this is exactly what is obtained as the light intensity passing through the arrangement is reduced to very low levels. A detector can be used to check that the photon passes through one – and only one – channel of the first crystal. The photon therefore appears to be 'aware' of the existence of the open horizontal channel, and is affected by it. Close the horizontal channel by inserting a stop between the

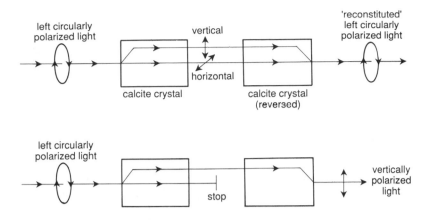

Fig. 2.6 Two calcite crystals placed 'back-to-back' produce some curious results.

crystals and the possibility of producing a left circularly polarized photon is lost: a vertically polarized photon emerges.

These two examples illustrate that, if we are right in our assumptions about the behaviour of individual photons, the further assumption that they pass through an apparatus as *localized* particles is wrong. The non-locality implied by the particle's dual wave–particle nature gives rise to effects that seem to contradict common sense. According to the orthodox interpretation of quantum theory, the state vector of a quantum particle is non-local: it 'senses' the entire measuring apparatus and can be affected by open slits or channels in polarization analysers in ways that a localized particle cannot. This is why the nature of the measuring apparatus is believed to be so important. The act of measurement itself — which begins with the blackening of photographic emulsion or the production of an electric current in a photomultiplier — 'concentrates' the state vector into a small region of space and hence 'localizes' the quantum particle.

Although we have tried to take care to preface many of the conclusions drawn in this chapter with the phase 'the orthodox interpretation', the fact is that this interpretation is the one taught to the majority of undergraduate chemists and physicists. It has therefore been important to go through it, if only so that readers can recognize it for what it is. If we are prepared to accept this interpretation, there are a number of consequences that follow automatically. Many scientists find these consequences so unacceptable that they reject the theory as somehow incomplete. This was Einstein's view. We will now examine these consequences in detail.

3
What does it mean?

3.1 POSITIVISM

The strange behaviour of photons described at the end of the last chapter immediately raises all sorts of questions about the meaning of quantum theory. We might be encouraged to look for this meaning by going back to the theory's mathematical structure: perhaps by trying to find out how we might better interpret some of its elements. However, it is a central argument of this book that, no matter where we look, we are always led back to philosophy. At first sight, this might seem to be an odd thing to claim. After all, modern textbooks on quantum physics and chemistry rarely (if ever) discuss philosophy. We accept that the behaviour of photons is strange, but surely it is something that we can at least study experimentally—do we need philosophy in these circumstances? But this is the whole point. Quantum theory directly challenges our understanding of the nature of the fundamental particles and the process of measurement, and we cannot go forward unless we adopt some kind of interpretation. As we will see, this interpretation has to be based on some philosophical position.

We will argue this point by first showing that the orthodox interpretation of quantum theory developed by Niels Bohr and his colleagues in Copenhagen (the one taught by design or default to most modern undergraduate scientists) is based on a particular philosophical outlook known as positivism. It will therefore be very helpful to begin by looking briefly at the positivists' line of reasoning. Do not be misled into thinking that arguments about philosophy are ultimately futile or irrelevant to important matters of concern to the experimental scientist. That this is not so will be amply demonstrated in Chapter 4.

Ernst Mach

Our first encounter with philosophy actually begins with a physicist. Ernst Mach was professor of physics at the Universities of Prague and Vienna from 1867 to 1901. Drawing on and extending a long philosophical tradition, he argued that scientific activity involves the study

of facts about nature revealed to us through our sensory perceptions
(perhaps aided by some instrument) and the attempt to understand their
interrelationship through observation and experiment. According to
Mach, this attempt should be made in the most economical way.

Mach rejected as non-scientific any statements made about the world
that are not empirically verifiable. What do we mean here by the word
empirical? The dictionary definition identifies empirical as purely experimental (i.e. without reference to theory) so that statements which are not
experimentally verifiable are rejected as non-scientific. However, this
definition does not seem to tell the whole story. Science is certainly not
about the mindless collecting of empirical facts about nature, it is about
interrelating those facts and making predictions on the basis of some
kind of theory. The key question concerns the way in which the concepts
of the theory should be interpreted.

Let us make use of a specific example. The philosophers of ancient
Greece developed a cosmological model of the universe which placed the
earth at the centre. An essential element in that model was the ideal of
the perfect circle, and the motions of the stars around the earth appear to
conform to this ideal. However, as seen from the earth, the motions
of the planets are far from circular. In about AD 150, the philosopher
Ptolemy attempted to explain the observed motions of the planets
around the earth by constructing an elaborate theory based on
epicycles — combinations of circles in which the ideal was at least preserved. To a certain extent he succeeded, but as observations became
more accurate he found that he had to add more and more epicycles.
Now Ptolemy's statements about the motions of the planets are empirically verifiable: if we use the theory in the prescribed manner, we
would expect to be able to compare them with observations and so
verify that they describe the motions of the planets (albeit with limited
accuracy).

In fact, we can easily imagine that we could develop a modern refine-
ment of the Ptolemaic system, with a very large number of epicycles, and
that with a little computer power we should be able to make some fairly
accurate predictions about the motions of the planets. Does this mean
that we should regard the epicycles to be 'real' in the sense that they represent real elements of the dynamics of planetary motion? Perhaps our
immediate reaction is to say 'Of course not!'. But why not? Ptolemy's
difficulties were created by his assumption that the sun and planets orbit
the earth, whereas a much more *economical* theory places the sun at the
centre of the solar system, as suggested by Copernicus. But does this
system necessarily represent reality any better than Ptolemy's?

Mach's point was that there is no purpose to be served by seeking to
describe a reality beyond our immediate senses. Instead, our judgement

should be guided by the criteria of verifiability (does the theory agree with experimental observations?) and simplicity (is it the simplest theory that will agree with the experimental observations?). Thus, if both the Ptolemaic and Copernican systems can be developed to the point that they make identical predictions for the motions of the planets, then our choice should be based solely on their relative conceptual or mathematical simplicity. In this case, the Copernican system wins out because it is the simplest.

In constructing a physical theory we should therefore seek the most economical way of organizing facts and making connections between them. We should not attach a deeper significance to the concepts used in a theory (such as epicycles) if they are not in themselves observable or subject to empirical verification. According to Mach, only those elements that we can perceive actually exist, and there is no point in searching for a physical reality that we cannot perceive: we can only know what we experience. Mach's criterion of what constituted a verifiable statement was particularly stringent. It led him to reject the concepts of absolute space and absolute time, and to side with Ludwig Boltzmann's opponents in rejecting the reality of atoms and molecules.

Speculations that are intrinsically not verifiable, that involve some kind of appeal to the emotions or to faith, are not scientific. However, these speculations, which are accomodated in a branch of philosophy called metaphysics, are not rejected outright. They are recognized as a legitimate part of the process of developing an attitude towards life, but they are perceived to have no place in science. This kind of approach is generally known as positivism.

Mach's views on space and time greatly influenced the young Einstein, whose admiration for Mach's work on mechanics never diminished. However, in his later life, Einstein had little time for Mach's positivist philosophy, and once stated that 'Mach was as good at mechanics as he was wretched at philosophy'.[†]

The Vienna Circle

Mach placed particular emphasis on the correct use of language, calling it 'the most wonderful economy of communication'. His views were enormously influential in the development of a new school of philosophical thought that emerged in Vienna in the early 1920s. Centred around Moritz Schlick, professor of philosophy at the University of Vienna, Rudolf Carnap, Otto Neurath and others, the 'Vienna Circle'

[†] Quotation from Pais, Abraham (1982). *Subtle is the Lord: the science and the life of Albert Einstein*. Oxford University Press.

extended their positivist outlook through the use of modern logic. They drew their inspiration from a wide variety of sources, particularly the work of the physicists Mach, Boltzmann and Einstein. Philosophically, their particular brand of positivism was foreshadowed in the work of the Scottish philosopher David Hume, and they were greatly influenced by the analytical approaches of their contemporaries Bertrand Russell in Cambridge and Ludwig Wittgenstein (a former student of Russell's) in Vienna.

The Vienna Circle began with the contention that the only true knowledge is scientific knowledge and that, in order to be meaningful, a scientific statement has to be a formally logical and verifiable statement. Their philosophy is sometimes known as logical positivism.

Scientists might think it rather obvious that science has to be logical. But the rigorous application of modern logic actually leads to an exhaustive analysis of the use of language and the meaning of words. This is necessary in order to rule out tautological or self-contradictory statements. At times, logical positivism appears more like philology than philosophy. Of course, we would never accept mathematical statements that use undefined terms or are self-contradictory: why should we expect less from language?

Most importantly, the use of logical analysis leads to the elimination of all metaphysical statements as meaningless. With one stroke, the logical positivists eliminated from philosophy centuries of 'pseudo-statements' about mind, being, reality and God, reassigning them to the arts alongside poetry and music. The views of the Vienna Circle came to dominate the philosophy of science in the middle of this century.

Positivism versus realism

Today, most modern scientists recognize that their activities involve dealing with observations, the results of experiments, and theoretical descriptions which reveal facts about nature. These facts are empirically verifiable and, indeed, are usually verified by other scientists. The scientists employ the methods of deductive logic in the formulation of theories to account for or explain the observed facts. The best (quite often the simplest and therefore most economical) theory is the one which accounts for all the known facts and can be used to make predictions, the accuracy of which can then be verified. Metaphysical speculations (about the existence of God, for example) do not usually play any part in the scientists' routine activity, although most scientists, when prompted, will certainly have a developed and highly distinctive metaphysical outlook that makes them complete as human beings.

Most Western scientists actually learn to use the *methods* of the positivist during their formal education. Although there are undoubtedly some 'grey' areas, young scientists are instructed by their teachers as to what qualifies as science, what 'doing science' means and how it should ideally be conducted. They learn to adopt a pragmatic, sceptical approach to science in which philosophy — and particularly metaphysics — appears to play no part.

However, for many scientists the stuff of their theories — atoms, electrons, photons, etc. — are quite 'real'. Many assume these objects to have an existence independent of the instruments used to produce the effects their theories are supposed to explain. It would, perhaps, be very difficult for high-energy physicists to justify the financial investments needed to build larger and larger particle accelerators if they were not convinced of the reality of the objects on which they wish to make measurements. Most scientists attempt to uncover the independent physical reality lying underneath the phenomena: to explain why the world is the way it is, which goes beyond merely registering the fact that instrument A will give effect B under conditions C. This position, in which it is held that there exists a reality which is independent of the observer and the instruments used to make observations, we will refer to as realism.

A rigorous positivist would question the usefulness of searching too hard for such an independent reality, although it would be a mistake to suppose that an uncompromising positivist stance would necessarily lead us to deny a reality that we cannot directly perceive. Schlick himself declared this kind of reasoning 'simply absurd', and refused to accept that it is implied in the philosophy of logical positivism (although we should note that not every logical positivist would necessarily agree with him).

However, there is a distinction to be made. A realist might be convinced that there is an independent reality 'out there' which is probed through observation and experiment. A positivist accepts that there are elements of an *empirical* reality which are probed in this way, but points out that the realist view involves a logical contradiction, since we have no way of observing an observer-independent reality and hence we cannot verify that such a reality exists. We have no means of acquiring knowledge of the physical world except through observation and experiment, and so the reality we probe is, of necessity, dependent on the observer for its existence. The positivist argues that, since we cannot verify the existence of an observer-independent reality, such a reality is metaphysical and therefore quite without meaning. The logical contradiction implied in the realist's view is side-stepped only by an appeal to the emotions or to faith.

A modern scientist might typically adopt the *methods* of the positivist but the *outlook* of the realist. If this position seems a little confusing and ill thought out, it is perhaps because scientists rarely spend time working out where they stand on these philosophical issues. Indeed, a pragmatic scientist might have little time for what seems like a kind of philosophical nit-picking. However, it is very difficult to avoid these issues in quantum theory. A quantum particle exhibits properties we associate with waves and particles. Its behaviour appears to be determined by the kind of instrument we use to probe its properties. One kind of instrument will tell us that the quantum particle is a wave. Another kind will tell us that it is a particle. All we can know is the *empirical* reality — here the quantum particle is a wave, here it is a particle. Is it therefore meaningful to speculate about what the quantum particle *is*?

Degrees of objectivity

Of all the virtues a scientist claims to uphold, objectivity is perhaps the most important. There are two ways of interpreting what we mean by 'objective'. In the first, we take the 'everyday' use of the word to imply that statements made by a scientist about experimental observations or measurements are ideally statements that do not depend on the scientist's personal motives, views, prejudices or religious belief.

In the second, we take the word objective to imply that there are no special circumstances that would lead us to expect that the scientist's observations are unique to that scientist. Using the information supplied, we expect to be able to repeat an experimental procedure and observe the same phenomena. In either meaning, the statements can be verified by others.

Of course, the practice falls somewhat short of the ideal. Scientists are people too, and are prone to all the failings we tend to associate with human nature. An otherwise objective scientist may defend an entrenched view (on a favourite theory, for example) long after overwhelming experimental evidence suggests that such views are logically indefensible. Scientists are also fallible: an experiment may be found to be unrepeatable because of special circumstances that pertained at the time the original observations were made, but which the scientist failed to communicate to others.

Nevertheless, many scientists are convinced that they pursue their chosen careers in an objective manner — they strive for the ideal. Furthermore, as we have argued, many believe that through their experiments they probe an underlying objective reality that is independent of them and their instruments. In their scientific papers they announce that this is the way the world is. Although they might use the positivist's methods, they are perhaps not prepared to accept the positivist's claim that their belief in an objective, independent reality is meaningless speculation.

For example, Einstein developed his special theory of relativity because he believed that reality, manifested in the laws of nature, should be completely objective, i.e. completely independent of the observer (Einstein was a realist). He achieved this by using mathematical relationships that made every inertial frame of reference equivalent — there is thus no special frame of reference unique to the observer. Out went the notions of absolute space and time.

But there is a weaker form of objectivity which we can identify with the positivist standpoint. In this view, we advocate an empirical reality which is not independent of the observer, but is the *same for all observers*. Even this may be in need for some qualification, some set of rules by which we make our judgements. It can be argued that Einstein's relativity is based on the need for weak objectivity and nothing more. Relativity places great emphasis on the central role of the observer and, in principle, says nothing about a reality which does not feature an observer. In fact, Einstein's whole approach provided much inspiration for the Vienna Circle. This distinction between the strong objectivity of the realist and the weaker objectivity of the positivist might seem to be subtle, but it captures the essence of the debate about the interpretation of quantum theory.

To summarize, we can identify two distinct philosophical positions — positivism and realism — which scientists tend to adopt (consciously or unconsciously) in their approach to their work. Scientists in both camps draw on the methods of deductive logic and make use of the criteria of verifiability and simplicity in the development of theories of the physical world. Both will strive for the ideal of objectivity in the way they apply these methods and criteria. However, for the positivist, the theory is merely an instrument which can be used to interrelate observed facts and make new predictions. It describes elements of an empirical reality which depends on the observer and the measuring device for its existence. This reality meets the demands of weak objectivity in the sense that it is the same for all observers. For the realist, the aim of a theory should be to describe an independent reality: it should describe how the world *is*. This reality meets all the demands of strong objectivity because it does not depend on the observer in any way.

3.2 THE COPENHAGEN INTERPRETATION

We saw in Chapter 1 that Schrödinger and Heisenberg adopted very different positions with regard to the interpretation of quantum theory.

Schrödinger was a realist: he believed that there is an underlying independent reality that his wave mechanics partly described. Heisenberg took a fairly uncompromising positivist stance, insisting that his matrix mechanics served its purpose as nothing more than an algorithm through which the results of experimental observations could be correlated and new predictions made. When Schrödinger demonstrated that the two approaches are mathematically equivalent, physicists were presented with a clear choice. This was more than just a choice between two equivalent mathematical formalisms: it was a choice between different philosophies.

Schrödinger's wave mechanics was the more popular, because of its instinctive (a positivist would say emotional or metaphysical) appeal. It held the promise that its further development might reveal a little more of that underlying independent reality, perhaps one of wave fields and their superpositions. In October 1926, Bohr invited Schrödinger to join with him and Heisenberg in Copenhagen to debate the issues, but Schrödinger remained unconvinced by their arguments. Their failure to persuade Schrödinger made Bohr and Heisenberg more determined than ever to find a radical new interpretation.

However, Bohr and Heisenberg themselves had different, deeply held views. As we have seen, they argued (sometimes bitterly) over the interpretation of quantum theory. In February 1927, Bohr departed to Norway for a skiing holiday, leaving Heisenberg in Copenhagen to marshall his thoughts and write his now famous paper on the uncertainty principle; a paper which he believed would completely demolish Schrödinger's wave field idea. When Bohr returned, he launched into Heisenberg's finished paper, treating it much like he would treat a first draft of one of his own papers. Heisenberg was dismayed: he wanted to publish his paper as quickly as possible to gain the upper hand in the debate with Schrödinger. Eventually, Wolfgang Pauli stepped in to referee the ensuing argument between Bohr and Heisenberg on the interpretation of the uncertainty principle. A consensus was reached, and Heisenberg added the footnote to his paper described on page 33.

These three physicists developed what became known as the Copenhagen interpretation of quantum theory. Its foundations are the uncertainty principle, wave–particle duality, Born's probabilistic interpretation of the wavefunction and the identification of eigenvalues as the measured values of observables. It is the interpretation which provides the basis of the postulates of quantum theory and the mathematical structure that results from them. It is an interpretation that is so well entrenched in physics that many students are surprised to discover that there are alternatives. We can now admit that the references to quantum theory's orthodox interpetation, made many times in Chapter 2, are actually references to the Copenhagen interpretation.

Bohr's philosophy

Bohr had developed his own distinctive philosophy even before he became a physicist. Interestingly, Bohr's emphasis was also on the use of language, and he is quoted as saying:[†]

Our task is to communicate experience and ideas to others. We must strive continually to extend the scope of our description, but in such a way that our messages do not thereby lose their objective or unambiguous character.

This sentiment was translated through to Bohr's scientific papers, the drafting of which would involve seemingly endless searching for just the right words or phrases that would communicate exactly what Bohr meant to say.

However, this emphasis on language went far beyond word-play. It transcended forms of written and verbal communication and included the sum of human experience. Bohr argued that we, as experimental scientists, design, perform, interpret and communicate the results of our experiments using the concepts of *classical* physics. We understand how large-scale laboratory instruments work only in terms of classical physics. The effect of an event occurring at the level of an individual quantum particle must be somehow amplified, or otherwise turned into some kind of macroscopic signal (such as a deflection of a pointer on a voltage scale) so that we can perceive and measure it. Our perceptions function at the level of classical physics and the only concepts with which we are entirely familiar, and for which we have a highly developed language, are classical concepts.

In his book *Physics and philosophy*, published in 1962, Heisenberg wrote that the Copenhagen interpretation of quantum theory actually rests on a paradox. This is the paradox of describing quantum phenomena in terms of idealized classical concepts. We only know of waves and particles — these are the concepts we have inherited from the experiences registered in our daily lives and from a long tradition of classical physics. This interpretation requires that we accept that we can never 'know' quantum concepts: they are simply beyond human experience and are therefore elements of an empirical reality. A quantum particle is neither a wave nor a particle. Instead we substitute the appropriate classical concept — wave or particle — as and when necessary.

[†] Petersen, Aage, in French, A. P. and Kennedy, P. J. (eds.) (1985). *Niels Bohr: a centenary volume*. Harvard University Press, Cambridge, MA.

Although it is unlikely that there was ever any significant interaction between the Copenhagen school of physicists and the Vienna Circle, their philosophies are in some respects quite compatible. Compare a typical Bohr statement[†]

There is no quantum world. There is only an abstract physical description. It is wrong to think that the task of physics is to find out how nature is. Physics concerns what we can say about nature.

with the following comment on logical positivism by the philosopher A. J. Ayer[‡]

The originality of the logical positivists lay in their making the impossibility of metaphysics depend not upon the nature of what could be known but upon the nature of what could be said.

We should take care not to place too much emphasis on this comparison, since there are areas in which the Copenhagen school and the Vienna Circle espoused quite different views. However, it is clear that Bohr shared some of the motives of the Vienna Circle in dismissing statements about an independent (and therefore metaphysical) reality as meaningless. He argued that we live in a classical world and our experiments are classical experiments. Go beyond these concepts and you cross the threshold between what you can know and what you cannot.

Complementarity

The Copenhagen interpretation requires that we consider very carefully the methods by which we acquire knowledge of the physical world. It shifts the focus of scientific activity from the objects of our studies to the relationships between those objects and the instruments we use to reveal their behaviour: the instrument takes centre-stage, alongside the object, and the distinction between them is blurred.

According to this interpretation, it is not meaningful to regard a quantum particle as having any intrinsic properties independent of some measuring instrument. Although we may speak of electron spin, velocity, orbital angular momentum, etc., these are properties we have assigned to an electron for convenience — each property becomes 'real' only when the electron interacts with an instrument specifically designed

---

[†] Petersen, Aage, in French, A.P. and Kennedy, P. J. (eds.) (1985). *Niels Bohr — a centenary volume*. Harvard University Press, Cambridge, MA.

[‡] Ayer, A. J. (ed.) (1959). *Logical positivism*. The Library of Philosophical Movements. The Free Press of Glencoe.

to reveal it. We may routinely use these concepts to predict how quantum particles will behave as though they were independent of ourselves and our instruments but ultimately we will need to test our predictions through experiment. Agreement between theory and experiment allows us to interpret these concepts as elements of an empirical reality. These concepts help us to correlate and describe our observations, but they have no meaning beyond their use as a means of connecting the object of our study with the instrument we use to study it.

Thus, when we make a statement such as 'This photon has vertical polarization', we should also make reference to (or at least be aware of) the experimental arrangement by which we have come by that knowledge. We might modify our statement thus: 'This photon was generated in such-and-such a way and was transmitted through a polarizing filter with its axis of maximum transmission oriented vertically with respect to some laboratory reference frame. Its passage through the filter was confirmed by the generation of a blackened spot on a piece of photographic film. This photon therefore combined with the instrument to reveal properties we associate with vertical polarization.' Note the emphasis on the past: in making the measurement the state of the photon was certainly changed irreversibly.

Bohr insisted that we can say nothing at all about a quantum particle without making very clear reference to the nature of the instrument which we use to make measurements on it. Thus, if our instrument is a double slit apparatus, and we study the passage of a photon through it, we know that we can understand the physics of the photon–instrument interaction using the wave concept as expressed in the photon's wave-function or state vector. If our instrument is a photomultiplier or a piece of photographic film, we know that the photon–instrument interaction can be understood in terms of a particle picture. We can design instruments to demonstrate a quantum particle's wave-like properties *or* its particle-like properties, but we cannot demonstrate both simultaneously. According to the Copenhagen interpretation, this is not because we lack the ingenuity to conceive of such an instrument, but because such an instrument is inconceivable.

As scientists, we perhaps find it difficult to resist the temptation to conjure up a mental picture of an individual photon existing in some kind of polarization state independently of our measurements. But according to the Copenhagen interpretation, such a mental picture would be at best unhelpful and at worst positively misleading.

Bohr summarized his views in a lecture delivered to a meeting of physicists on 16 September 1927 at Lake Como in Italy. It was during this lecture that he introduced his idea of complementarity. This idea went through many refinements and restatements, but now tends to be presented in terms of wave–particle duality. Bohr argued that although the wave picture and the particle picture are mutually exclusive, they are not contradictory, but complementary. For Bohr, complementarity lay at the heart of the strange nature of the quantum world. The uncertainty principle becomes merely a mathematical statement expressing the limits imposed on our ability to make measurements based on complementary concepts of classical physics. The mathematical formalism of quantum theory becomes an attempt to repackage complementary wave and particle descriptions in a single, all-encompassing theory. This does not imply that the theory is wrong or somehow incomplete. On the contrary, it is the best we can do and goes as far as we can go.

Complementarity and objectivity

Because of the emphasis placed on the importance of the observer or observing instrument, many physicists and philosophers have accused the Copenhagen interpretation of being subjective. Clearly, the subject (the observer) appears to exercise remarkable powers over reality, with the freedom to choose what kind of reality is to be probed. In the language of quantum measurement described in Chapter 2, a simple re-orientation of a polarizing filter changes instantaneously the measurement eigenstates of a quantum system, thereby changing the nature of the reality that can be exposed. The whole process of expanding the state vector in terms of the measurement eigenstates then becomes a subjective process — what is written down depends upon the subject's personal preferences, not on the independent, objectively real properties of the object under study.

This charge of subjectivism is unfair. In many of his most oft-quoted statements, Bohr insisted that he was searching for objectivity. But his was the weaker objectivity that we have in this book associated with positivism rather than the strong objectivity of the realist. The state vector of the Copenhagen interpretation might not reflect an objectively real behaviour, but the information communicated by one physicist to another about the methods used to analyse the state vector in terms of the measuring instrument (using the language of classical physics) means that the experiment can be repeated, the experiences shared and the interpretation understood.

The complementarity of object and subject (the instrument or the observer) is as important in the Copenhagen interpretation as any other form of complementarity. As we discussed in Section 2.6, the state vector describes the evolution of a quantum system in a way that is quite deterministic. Once the initial conditions have been established, the future behaviour of a quantum particle is predictable through the quantum mechanical laws of motion. However, this determinism does not apply to our classical conception of space-time. It is not a determinism in any 'normal', classical sense of the word. In order to deal in a practical way with the state vector it has to be projected into a form that we can recognize within the reference frame provided by classical physics — we must make a measurement on it. But the act of measurement destroys the continuity provided by the state vector. The space-time description and the probabilistic description in terms of the state vector are complementary.

Heisenberg again:[†]

Our actual situation in research work in atomic physics is usually this: we wish to understand a certain phenomenon, we wish to recognise how this phenomenon follows from the general laws of nature. Therefore, that part of matter or radiation which takes part in the phenomenon is the natural 'object' in the theoretical treatment and should be separated in this respect from the tools used to study the phenomenon. This again emphasises a subjective element in the description of atomic events, since the measuring device has been constructed by the observer, and we have to remember that what we observe is not nature in itself but nature exposed to our method of questioning.

This quotation nicely captures the positivist flavour of the Copenhagen interpretation.

Criticisms

We should recognize what we are dealing with here. The Copenhagen interpretation essentially states that in quantum theory we have reached the limit of what we can know. To try to go beyond this limit is pointless (how can we know something that is unknowable?). The argument is that any attempt to introduce a new concept to describe an underlying independent reality inevitably involves a reworking of familiar classical concepts. We always return to the idealized concepts that summarize the fullest extent of our knowledge — waves and particles.

It is interesting to note that in some branches of physics, scientists have long since given up making such attempts. For example, quarks are now generally accepted as one of the families of fundamental constituents of matter and are therefore accepted as elements of an empirical reality. They are categorized according to their properties in various 'flavours', termed up, down, strange, charm, bottom and top. Experimental evidence supporting the 'existence' of the first five types of quark has been obtained, but the top quark has so far remained elusive. The names given to these particles are intentionally abstract: they are intended only

[†] Heisenberg, Werner, (1989). *Physics and philosophy*. Penguin, London.

to provide an economical means of communicating their properties and status in a somewhat abstract theory. This is Bohr's philosophy writ large.

But should we accept that we have reached the end of the road? A charge frequently levelled at positivism is that it is sterile; it denies that there is a way forward through just the kind of metaphysical speculation that can introduce concepts which begin life as abstract mathematical constructions (such as atoms and quarks) into elements of reality. Can we afford not to push science along apparently 'meaningless' paths? What if, despite appearances, we have not reached the limit after all? What if there *is* something more to discover about reality if only we have the wit to ask the right questions? Whatever our personal thoughts on this matter, we should admit that it goes against the grain of human nature not to *try*.

The Copenhagen school of physicists was convinced that its interpretation of quantum theory was the only sensible interpretation. Other physicists disagreed, however. As we have seen, Schrödinger refused to bow to pressure from Bohr and Heisenberg to reconsider his position. Einstein was never comfortable with quantum theory's implications for causality (another classical concept, the Copenhagen school would quickly point out). These two were only the most eminent and directly involved of the physicists who were unhappy with what the Copenhagen school was saying. Einstein in particular confronted Bohr head on in a now famous debate on the meaning of quantum theory.

3.3 THE BOHR–EINSTEIN DEBATE

Although invited, Einstein did not attend the meeting of physicists at Lake Como in September 1927 at which Bohr first presented his ideas about complementarity. He was nevertheless very active in the debate. It appears that he had had some earlier correspondence with Heisenberg concerning the uncertainty principle, the details of which had appeared in Heisenberg's paper published in March. Einstein probably expressed once again his worries about the principle's implications for strict causality. At the same time, he was developing his own ideas about the interpretation of quantum theory, based on the statistical properties of large collections of particles.

Finally, on 24 October 1927, Einstein, Bohr and many other leading physicists assembled in Brussels for the fifth Solvay Conference, on 'Electrons and Photons'.

The fifth Solvay Conference

Einstein did not present a paper at the conference, and made little contribution to the formal proceedings. However, the discussion on those comments he did make spilled over into the dining room of the hotel in which all the conference participants were staying and which became the scene of one of the most important scientific debates ever witnessed, as Einstein directly challenged Bohr over the meaning of quantum theory. At stake was the interpretation of quantum theory and its implications for the way we attempt to understand the physical world. The outcome would determine the directions of the future development of quantum physics.

This debate has been described in great detail by Bohr himself in a book published in 1949 in celebration of Einstein's 70th birthday. Einstein began by expressing his general reservations about quantum theory by reference to an experiment involving the diffraction of a beam of electrons or photons through a narrow aperture, as shown in Fig. 3.1. The diffraction pattern appears on a second screen and is recorded (using photographic film, say). According to the Copenhagen interpretation, the behaviour of each individual quantum particle is described by an appropriate wavefunction and it is the properties of the wavefuntion that give rise to the diffraction pattern. However, at the moment the wavefunction impinges on the second screen, it 'collapses' instantaneously, producing a localized spot on the screen which indicates 'a particle struck here'.

Einstein objected to this way of looking at the process. Suppose, he said, that the particle is observed to arrive at position A on the second screen (see Fig. 3.1). In making this observation, we learn not only that

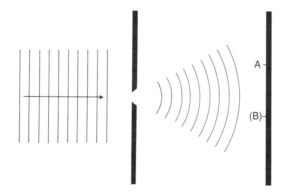

Fig. 3.1 The simple electron or photon diffraction experiment cited by Einstein in his debate with Bohr. Reprinted from The Library of Living Philosophers, Vol. VII, *Albert Einstein: philosopher-scientist*, edited by Paul Arthur Schilpp, by permission of the publisher (La Salle, IL: Open Court Publishing Company, 1949), p. 212.

the particle arrived at A, but also that it definitely did not arrive at position B. What is more, we learn of the particle's non-arrival at B instantaneously with the observation of its arrival at A. Before observation, the probability of finding the particle is, supposedly, 'smeared out' over the whole screen.

Einstein believed that the collapse of the wavefunction implies a peculiar 'action at a distance'. The particle, which is somehow distributed over a large region of space, becomes localized instantaneously, the act of measurement appearing to change the physical state of the system far from the point where the measurement is actually made. Einstein felt that this kind of action at a distance violated the postulates of special relativity.

There is an alternative description, however. What if the wavefunction represents a probability amplitude not for a single quantum particle, but for a large collection (called an ensemble) of particles which is described in terms of a single wavefunction? According to this view, each individual particle passes through the aperture along a defined, localized path, to arrive at the second screen. There are many such paths possible, and the diffraction pattern thus reflects the statistical distribution of large numbers of particles each following different but defined paths. This distribution is related to the modulus-squared of the wavefunction, which expresses the probability density of one (of many) particles rather than a probability density for each individual particle.

We should note that we cannot choose between these possibilities by observing what happens to an individual quantum particle. Both descriptions say that one particle passes through the aperture to arrive at one specific location on the second screen. In the first description, the point of arrival is determined at the moment the particle interacts with the detector, with a probability given by $|\psi|^2$. In the second description, the point of arrival is determined by the actual path which the particle follows, which is in turn obtained from a statistical probability given by $|\psi|^2$. In both cases we see the diffraction pattern only when we have detected a large number of particles.

Thought experiments

Einstein then attacked the Copenhagen interpretation of quantum theory by attempting to show that it is inconsistent. The debate took the form of a series of puzzles, developed by Einstein as hypothetical experiments. These 'thought' experiments were not intended to be taken too literally as practical experiments that could be carried out in the laboratory. It was enough for Einstein that the experiments could be conceived and carried out *in principle*.

Einstein asked the assembled audience what might happen if a quantum particle passed through an apparatus such as the one shown in Fig. 3.1 under conditions where the transfer of momentum between the particle and the first screen is carefully controlled and observed. A particle hitting the screen as it passes through the aperture would be deflected, its path beyond being determined by the conservation of momentum. Now imagine that we insert another screen — one with two slits — between the first screen and the detector (Fig. 3.2). If we control the transfer of momentum between the particle and the first screen, we should be able to discover towards which slit in the second screen the particle is deflected. If the particle is ultimately detected, we can deduce from our measurements that it passed through one or other of the two slits, and we have thus determined the particle's trajectory through the apparatus. We can now leave the apparatus to detect a large number of particles — one after the other — from which we expect to see a double-slit interference pattern. Thus, Einstein concluded, we can demonstrate the particle-like (defined trajectory) and wave-like (interference) properties of quantum particles simultaneously, in contradiction to Bohr's complementarity idea, proving that the Copenhagen interpretation is inconsistent.

Bohr's reaction was to take the thought experiment a stage further. He sketched out in a pseudo-realistic style the kind of apparatus that would

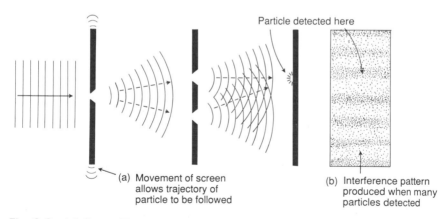

Particle detected here

(a) Movement of screen
allows trajectory of
particle to be followed

(b) Interference pattern
produced when many
particles detected

Fig. 3.2 (a) Controlling and observing the momentum transferred between a quantum particle and the first screen allows the trajectory of the particle to be traced through a double-slit apparatus. (b) After many particles have passed through the apparatus, the double-slit interference pattern should be visible. Adapted from The Library of Living Philosophers, Vol. VII, *Albert Einstein: philosopher-scientist*, edited by Paul Arthur Schilpp, by permission of the publisher (La Salle IL: Open Court Publishing Company, 1949), p. 216.

be needed to make the measurements to which Einstein referred. His purpose was not to try to imagine how the experiments could be done in practice, but primarily to focus on what he saw to be flaws in Einstein's arguments.

Thus, controlling and observing the transfer of momentum from the quantum particle to the first screen requires that the screen be capable of movement in the vertical plane. Observing the recoil of the screen in one direction or the other as the particle passed through the aperture would then allow the experimenter to draw conclusions about the direction in which the particle had been deflected. Bohr envisaged a screen suspended by two weak springs, as shown in Fig. 3.3. A pointer and scale inscribed on the screen allows the measurement of the amount of movement of the screen, and hence the momentum imparted to it by the particle. The fact that Bohr had in mind a macroscopic apparatus presents no problem, provided we assume that the apparatus is sufficiently sensitive to allow observation of individual quantum events. This sensitivity is important, as we will see below.

Bohr had to demonstrate the consistency of the uncertainty principle,

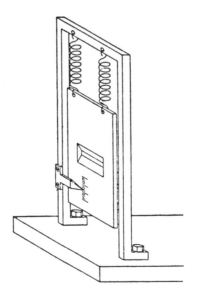

Fig. 3.3 Hypothetical instrument designed by Bohr to demonstrate how the measurement of the momentum transfer to the first screen might be made. Reprinted from The Library of Living Philosophers, Vol. VII, *Albert Einstein: philosopher-scientist*, edited by Paul Arthur Schilpp, by permission of the publisher (La Salle, IL: Open Court Publishing Company, 1949), p. 220.

and hence of the complementarity idea, when applied to the analysis of this kind of thought experiment. He argued that controlling the transfer of momentum to the screen in the way Einstein suggested *must* imply a concomitant uncertainty in the screen's position in accordance with the uncertainty principle. If we measure the screen's momentum in the vertical plane with a precision Δp_x, an uncertainty $\Delta x \geqslant h/4\pi\Delta p_x$ in the position of the screen must result.

Bohr was able to show that the resulting uncertainty Δx in the position of the aperture in the first screen corresponds approximately to the distance between adjacent fringes in the double-slit interference pattern. The positions of the fringes are therefore uncertain by an amount equal to the spacing between them and the interference pattern is 'washed out'. Controlling the transfer of momentum from the particle to the first screen allows us to follow the trajectory of the particle through the apparatus, but prevents us from observing interference effects, in accordance with the complementary nature of wave and particle properties.

Bohr's argument rests on the assumption that controlling and measuring the momentum transferred to the first screen sufficiently precisely to determine the particle's future direction automatically leads to an uncertainty in the screen's position. Why should this be? Bohr's answer was that, in order to read the scale inscribed on the first screen sufficiently accurately, it has to be illuminated. This illumination involves the scattering of photons from the screen and hence an uncontrollable transfer of momentum, preventing the momentum transfer from the quantum particle to be measured precisely. We can only measure the latter with precision if we reduce the illumination completely, but then we cannot determine the position of the pointer against the scale. Bohr concluded:[†]

. . . we are presented with a choice of *either* tracing the path of a particle *or* observing interference effects, which allows us to escape from the paradoxical necessity of concluding that the behaviour of an electron or a photon should depend on the presence of a slit in the [second screen] through which it could be proved not to pass. We have here to do with a typical example of how the complementary phenomena appear under mutually exclusive experimental arrangements and are just faced with the impossibility, in the analysis of quantum effects, of drawing any sharp separation between an independent behaviour of atomic objects and their interaction with the measuring instruments which serve to define the conditions under which the phenomena occur.

Einstein did not give up. He produced further thought experiments that we do not have room to consider fully here. He could not shake his deeply felt misgivings about the Copenhagen interpretation and forced

---

[†] Bohr, N. in Schilpp, P. A. (ed.) (1949). *Albert Einstein: philosopher-scientist*. The Library of Living Philosophers, Open Court Publishing Company, La Salle, IL.

Bohr to defend it. The fifth Solvay Conference ended with Bohr having successfully argued for the logical consistency of the Copenhagen interpretation, but he had failed to convince Einstein that it was the only interpretation.

The photon box experiment

The debate recommenced at the sixth Solvay Conference, which was held in Brussels between 20–25 October 1930. Although the conference was devoted to the physics of magnetism, there was keen interest in the discussions on the interpretation of quantum theory that took place between the conference's formal proceedings. Einstein described his latest and most ingenious thought experiment, a further development of one that he had originally used in discussions at the fifth Solvay Conference. This is the 'photon box' experiment.

Suppose, said Einstein, that we build an apparatus consisting of a box which contains a clock mechanism connected to a shutter. The shutter closes a small hole in the box. We fill the box with photons and weigh it. At a predetermined and precisely known time, the clock mechanism triggers the opening of the shutter for a very short time interval and a single photon escapes from the box. The shutter closes. We reweigh the box and, from the mass difference and special relativity ($E = mc^2$) we determine the precise energy of the photon that escaped. By this means, we have measured precisely the energy and time of passage of a photon through a small hole, in contradiction to the energy–time uncertainty relation.

Bohr's immediate reaction has been described by Léon Rosenfeld:[†]

During the whole evening he was extremely unhappy, going from one to the other and trying to persuade them that it couldn't be true, that it would be the end of physics if Einstein were right; but he couldn't produce any refutation.

Bohr experienced a sleepless night, searching for the flaw in Einstein's argument that he was convinced must exist. By breakfast the following morning he had his answer.

Again Bohr produced a sketch of the apparatus that would be required to make the measurements in the way Einstein had described them, and this is shown in Fig. 3.4. The whole box is suspended by a spring and fitted with a pointer so that its position can be read on a scale affixed to the support. A small weight is added to bring the pointer to the zero on the scale. The clock mechanism is shown inside the box, connected to the shutter. After the release of one photon, the small weight is replaced by another, heavier weight so that the pointer is returned to the zero of the

---

[†] Rosenfeld, L. (1968) in *Proceedings of the fourteenth Solvay conference*. Interscience, NY.

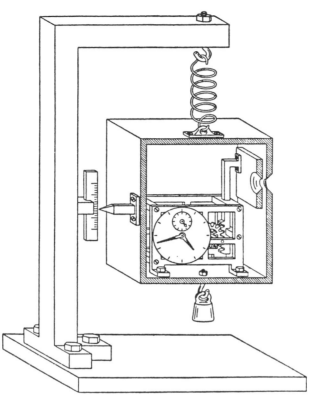

Fig. 3.4 The photon box experiment. Hypothetical instrument designed by Bohr to show how the measurements suggested by Einstein might be carried out. Reprinted from The Library of Living Philosophers, Vol. VII, *Albert Einstein: philosopher-scientist*, edited by Paul Arthur Schilpp, by permission of the publisher (La Salle, IL: Open Court Publishing Company, 1949), p. 227.

scale. The weight required to do this can be determined independently with arbitrary precision. The difference in the two weights required to balance the box gives the mass lost through the emission of one photon, and hence the energy of the photon.

Let us focus on the first weighing, before the photon escapes. Obviously, we will have set the clock mechanism to trigger the shutter at some predetermined time and the box will be sealed. The actual reading of the clock face is, of course, not possible since this would involve an exchange of photons — and hence energy — between the box and the outside world. To weigh the box, we must select a weight that just sets the pointer to the zero of the scale. However, to make a precise position measurement, the pointer and scale will again need to be illuminated and, following Bohr's earlier arguments, this implies an uncertainty in the momentum of the box. How does this affect the weighing? The uncontrollable transfer of momentum to the box causes it to jump about unpredictably. Although we can fix the box's instantaneous position against the scale, the sizeable interaction during the act of measurement means that the box will not stay in that position. Bohr argued that we can increase the precision of measurement of the *average* position by allowing ourselves a long time interval in which to perform the whole balancing procedure. This will give us the necessary precision in the weight of the box. Since we can anticipate the need for this, we can set the clock mechanism so that it opens the shutter after the balancing procedure has been completed.

Now comes Bohr's coup de grâce. According to Einstein's general theory of relativity, the rate of a clock moving in a gravitational field changes, and so the very act of weighing a clock effectively changes the way it keeps time. This phenomenon is responsible for the red shift in the frequency of radiation emitted from the sun and stars. Because the box is jumping about unpredictably in a gravitational field (owing to the act of measuring the position of the pointer), the rate of the clock is changed in a similarly unpredictable manner. This introduces an uncertainty in the exact timing of the opening of the shutter which depends on the length of time needed to weigh the box. The longer we make the balancing procedure (the greater the ultimate precision in the measurement of the energy of the photon), the greater the uncertainty in its exact moment of release. Bohr was able to show that the relationship between the uncertainties of energy and time is in accord with the uncertainty principle. This response was hailed as a triumph for Bohr and for the Copenhagen interpretation of quantum theory. Einstein's own general theory of relativity had been used against him.

However, Einstein remained stubbornly unconvinced, although he did change the nature of his attacks on the theory. Instead of arguing that the theory is inconsistent, he began to develop arguments that he believed demonstrated its *incompleteness*. When discussing the photon box experiment, Einstein conceded that it now appeared to be 'free of contradictions', but in his view it still contained 'a certain unreasonableness'.

We should not leave the photon box experiment without noting that many physicists, including Bohr, have since examined it over again in considerable detail. Some have rejected Bohr's response completely, denying that the uncertainty principle can be 'saved' in the way Bohr maintained. Others have rejected Bohr's response but have given alternative reasons why the uncertainty principle is not invalidated. Despite these counterproposals, the prevailing view in the physics community at the time appears to be that Bohr won this particular round in his debate with Einstein. However, Bohr appears to have been quite unprepared for Einstein's next move.

3.4 IS QUANTUM MECHANICS COMPLETE?

In May 1935, Einstein published a paper in the journal *Physical Review* co-authored with Boris Podolsky and Nathan Rosen. This paper is entitled: 'Can quantum-mechanical description of physical reality be considered complete?', and the abstract reads as follows:[†]

In a complete theory there is an element corresponding to each element of reality. A sufficient condition for the reality of a physical quantity is the possibility of predicting it with certainty, without disturbing the system. In quantum mechanics in the case of two physical quantities described by non-commuting operators, the knowledge of one precludes the knowledge of the other. Then either (1) the description of reality given by the wave function in quantum mechanics is not complete or (2) these two quantities cannot have simultaneous reality. Consideration of the problem of making predictions concerning a system on the basis of measurements made on another system that had previously interacted with it leads to the result that if (1) is false then (2) is also false. One is thus led to conclude that the description of reality as given by a wave function is not complete.

The Einstein–Podolsky–Rosen (EPR) thought experiment involves measurements made on one of two quantum particles that have somehow interacted and moved apart. We will denote these as particle A and particle B. The position q_A and momentum p_A of particle A are complementary observables and we cannot measure one without introducing an uncertainty in the other in accordance with Heisenberg's uncertainty principle. Similar arguments can be made for the properties q_B and p_B of particle B.

Now consider the quantities $Q = q_A - q_B$ and $\hat{P} = \hat{p}_A + \hat{p}_B$, where $\hat{p}_A = -i\hbar\partial/\partial q_A$ and $\hat{p}_B = -i\hbar\partial/\partial q_B$. The commutator $[Q, \hat{P}]$ is given by

$$[Q,\hat{P}] = Q\hat{P} - \hat{P}Q = (q_A - q_B)(\hat{p}_A + \hat{p}_B) - (\hat{p}_A + \hat{p}_B)(q_A - q_B)$$

$$= q_A\hat{p}_A + q_A\hat{p}_B - q_B\hat{p}_A - q_B\hat{p}_B - (\hat{p}_A q_A - \hat{p}_A q_B + \hat{p}_B q_A - \hat{p}_B q_B)$$

$$= (q_A\hat{p}_A - \hat{p}_A q_A) + (q_A\hat{p}_B - \hat{p}_B q_A) - (q_B\hat{p}_A - \hat{p}_A q_B)$$

$$- (q_B\hat{p}_B - \hat{p}_B q_B)$$

$$= [q_A,\hat{p}_A] + [q_A,\hat{p}_B] - [q_B,\hat{p}_A] - [q_B,\hat{p}_B] \qquad (3.1)$$

[†] Einstein, A., Podolsky, B. and Rosen, N. (1935). *Physical Review.* **47**, 777.

In this equation, $[q_A, \hat{p}_A] = [q_B, \hat{p}_B] = i\hbar$ (the position–momentum commutation relation) and $[q_A, \hat{p}_B] = [q_B, \hat{p}_A] = 0$, since these operators refer to different quantum particles. Hence, $[Q, \hat{P}] = 0$, the operators Q and $\hat{P}$ commute and there is no restriction on the precision with which we can measure the difference between the positions of particles A and B and the sum of their momenta.

A reasonable definition of reality

EPR allowed themselves what seems at first sight to be a fairly reasonable definition of physical reality:[†]

If, without in any way disturbing a system, we can predict with certainty (i.e. with a probability equal to unity) the value of a physical quantity, then there exists an element of physical reality corresponding to this physical quantity.

The purpose of this statement is to make clear that for each particle considered individually, the measurement of one physical quantity (the position of B, say) with certainty ($\Delta q_B = 0$) implies an infinite uncertainty in its momentum (since $\Delta p_B \geqslant h/4\pi \Delta q_B$). Therefore, according to EPR's definition of reality, under these circumstances the position of particle B is an element of physical reality but the momentum is not. Obviously, by choosing to perform a different measurement, we can establish the reality of the momentum of particle B but not its position. The Copenhagen interpretation of quantum theory insists that we can establish the reality of one or the other of two complementary physical quantities but not both simultaneously.

But we have shown above that the difference in the positions of particles A and B and the sum of their momenta are quantities whose operators commute. The Copenhagen interpretation says that we can therefore establish the physical reality of these quantities simultaneously. It is enough for the EPR argument that these quantities are simultaneously real *in principle*, although their actual determination might require a physical measurement.

Now suppose we allow the two particles to interact and move a long distance apart. We perform an experiment on particle A to measure its position with certainty. We know that $(q_A - q_B)$ must be a physically real quantity and so we can in principle deduce the position of particle B also with certainty. We therefore conclude that q_B must be an element of physical reality according to the EPR definition. However, suppose instead that we choose to measure the momentum of particle A with

[†] Einstein, A., Podolsky, B. and Rosen, N. (1935). *Physical Review.* **47**, 777.

certainty. We know that $(p_A + p_B)$ must be physically real, and so we can in principle deduce the momentum of particle B with certainty. We conclude that it too must be an element of physical reality. Thus, although we have not performed any measurements on particle B following its separation from A, we can, in principle, establish the reality of either its position or its momentum from measurements we *choose* to perform on A which, by definition, do not disturb B.

The Copenhagen interpretation denies that we can do this. We are forced to accept that if this interpretation of quantum theory is correct, the physical reality of *either* the position *or* momentum of particle B is determined by the nature of the measurement we choose to make on a completely different particle an arbitrarily long distance away. EPR argued that 'No reasonable definition of reality could be expected to permit this.'

As presented above, the EPR argument is based on a hypothetical experiment and is concerned with matters of principle. At the time the argument was developed, it was unimportant that the proposed experiment is difficult, if not impossible, to perform. However, we will see in the next chapter that the experimental study of the behaviour of quantum particles that have interacted and moved apart is made much more practicable if their spin properties are probed rather than their positions and momenta.

Spooky action at a distance

The EPR thought experiment strikes right at the heart of the Copenhagen interpretation. If the uncertainty principle applies to an individual quantum particle, then it appears that we must invoke some kind of action at a distance if the reality of the position or momentum of particle B is to be determined by measurements we choose to perform on A.

Whether it involves a change in the physical state of the system or merely some kind of communication, the fact that this action at a distance must be exerted instantaneously on a particle an arbitrarily long distance away from our measuring device suggests that it violates the postulates of special relativity, which restricts any signal to be communicated no faster than the speed of light. EPR did not believe that such action at a distance is necessary: the position and momentum of particle B are defined all along and, as there is nothing in the wavefunction which tells us how these quantities are defined, quantum theory is incomplete. EPR concluded:[†]

[†] Einstein, A., Podolsky, B. and Rosen, N. (1935). *Physical Review.* 47, 777.

While we have thus shown that the wave function does not provide a complete description of physical reality, we left open the question of whether or not such a description exists. We believe, however, that such a theory is possible.

Bohr's reply

Bohr first heard of the EPR argument from Léon Rosenfeld, who was at that time working with Bohr in Copenhagen. Rosenfeld later reported that:[†]

. . . this onslaught came down upon us like a bolt from the blue. Its effect on Bohr was remarkable . . . as soon as Bohr had heard my report of Einstein's argument, everything else was abandoned: we have to clear up such a misunderstanding at once. We should reply by taking up the same example and showing the right way to speak about it. In great excitement, Bohr immediately started dictating to me the outline of such a reply. Very soon, however, he became hesitant. 'No, this won't do, we must try all over again . . . we must make it quite clear.' So it went on for a while, with growing wonder at the unexpected subtlety of the argument.

Bohr's reply to the EPR argument was published in *Physical Review* in October 1935. He chose to use the same title that EPR had used in May and the abstract reads as follows:[‡]

It is shown that a certain 'criterion of physical reality' formulated in a recent article with the above title by A. Einstein, B. Podolsky and N. Rosen contains an essential ambiguity when it is applied to quantum phenomena. In this connection a viewpoint termed 'complementarity' is explained from which quantum-mechanical description of physical phenomena would seem to fulfill, within its scope, all rational demands of completeness.

Bohr's paper is essentially a summary of the complementarity idea and its application to quantum theory. He rejects the argument that the EPR thought experiment creates serious difficulties for the Copenhagen interpretation and stresses once again the importance of taking into account the necessary interactions between the objects of study and the measuring devices. He wrote:

From our point of view we now see that the wording of the above-mentioned criterion of physical reality proposed by Einstein, Podolsky and Rosen contains an ambiguity as regards the meaning of the expression 'without in any way disturbing a system' . . . there is essentially the question of *an influence on the*

[†] Rosenfeld, L. in Rozenthal, S. (1967). *Niels Bohr; his life and work as seen by his friends and colleagues*. North-Holland, Amsterdam.
[‡] Bohr, N. (1935). *Physical Review*. **48**, 696.

very conditions which define the possible types of predictions regarding the future behaviour of the system. Since these conditions constitute an inherent element of the description of any phenomenon to which the term 'physical reality' can be properly attached, we see that the argumentation of the mentioned authors does not justify their conclusion that quantum-mechanical description is essentially incomplete.

Many in the physics community seemed to accept that Bohr's paper put the record straight on the EPR experiment. I find Bohr's wording really rather vague and unconvincing. His emphasis is once again on the important role of the measuring instrument in *defining* the elements of reality that we can observe. Thus, setting up an apparatus to measure the position of particle A with certainty, from which we can infer the position of particle B, *excludes* the possibility of measuring the momentum of A and hence inferring the momentum of B. If there is no mechanical disturbance of particle B (as EPR assume), its elements of physical reality must be defined by the nature of the measuring device we have selected for use with particle A.

Does this necessarily imply an action at a distance? Certainly, if we could somehow delay our choice of measuring instrument (position versus momentum) until almost the last moment, then in principle the information available to us about a particle some considerable distance away changes instantaneously. An action at a distance will be required if the measurement performed on A changes the physical state of B or results in some kind of communication to B of particle A's changed circumstances.

If the physical state of both particles is described by a single wavefunction, which would be the case for two particles that have interacted, then the measurement collapses the wavefunction into one of the measurement eigenfunctions, as described in Section 2.6. The changes in the wavefunction must be felt through the whole of the quantum system, including particle B, even though it may by that time have travelled halfway across the universe.

Now if the wavefunction reflects only our state of knowledge of the quantum system, then its collapse would not seem to affect the system's physical properties. However, the problem remains that the collapse of the wavefunction requires that those physical properties become manifest in the quantum system where before they were not defined. The physical properties of particle B suddenly become real, where before they were not. It is difficult to imagine how this might happen without some kind of change in the physical state of a distant particle.

Einstein separability

In June 1935, Schrödinger wrote to congratulate Einstein on the EPR paper. He wrote:[†]

I was very happy that in the paper just published in [*Physical Review*] you have evidently caught dogmatic [quantum mechanics] by the coat-tails . . . My interpretation is that we do not have a [quantum mechanics] that is consistent with relativity theory, i.e, with a finite transmission speed of all influences. We have only the analogy of the old absolute mechanics . . . The separation process is not at all encompassed by the orthodox scheme.

Schrödinger's reference to the 'separation process' highlights the essential difficulty that the EPR argument creates for the Copenhagen interpretation. According to this interpretation, the wavefunction for the two-particle quantum state does not separate as the particles themselves separate in space–time. Instead of dissolving into two completely separate wavefunctions, one associated with each particle, the wavefunction is 'stretched' out and, when a measurement is made, collapses instantaneously despite the fact that it may be spread out over a large distance.

EPR's definition of physical reality requires that the two particles are considered to be isolated from each other, i.e. they are no longer described by a single wavefunction at the moment a measurement is made. The reality thus referred to is sometimes called 'local reality' and the ability of the particles to separate into two locally real independent physical entities is sometimes referred to as 'Einstein separability'. Under the circumstances of the EPR thought experiment, the Copenhagen interpretation denies that the two particles are Einstein separable and therefore denies that they can be considered to be locally real (at least, before a measurement is made on one or other of the particles, at which point they both become localized).

Entangled states and Schrödinger's cat

Motivated largely by the EPR paper, Schrödinger published in 1935 details of one of the most famous of the paradoxes of quantum theory, derived from one of the most difficult conceptual problems associated with quantum measurement. In our discussion of this topic in Chapter 2, the notion of the collapse of the wavefunction was presented without reference to the point in the measurement process at which the collapse occurs. Readers might have assumed that the collapse occurs at the

---

[†] Schrödinger, Erwin, letter to Einstein, Albert, 7 June 1935.

moment the microscopic quantum system interacts with the macroscopic measuring device. But is this assumption justified? After all, a macroscopic measuring device is composed of microscopic entities — molecules, atoms, protons, neutrons and electrons. We could argue that the interaction takes place on a microscopic level and should, therefore, be treated using quantum mechanics.

Suppose a quantum system described by some state vector $|\Psi\rangle$ interacts with a measuring instrument whose measurement eigenstates are $|\psi_+\rangle$ and $|\psi_-\rangle$. These eigenstates combine with the macroscopic instrument to reveal one or other of the two possible outcomes, which we can imagine to involve the deflection of a pointer either to the left ($+$ result) or the right ($-$ result). Recognizing that the instrument itself consists of quantum particles, we describe the state of the instrument before the measurement in terms of a state vector $|\phi_0\rangle$, corresponding to the central pointer position. The total state of the quantum system plus the measuring instrument before the measurement is made is described by the state vector $|\Phi_0\rangle$, which is given by the product:

$$|\Phi_0\rangle = |\Psi\rangle|\phi_0\rangle = \frac{1}{\sqrt{2}}\left[|\psi_+\rangle + |\psi_-\rangle\right]|\phi_0\rangle$$
$$= \frac{1}{\sqrt{2}}\left[|\psi_+\rangle|\phi_0\rangle + |\psi_-\rangle|\phi_0\rangle\right] \tag{3.2}$$

where we have made use of the expansion theorem to express $|\Psi\rangle$ in terms of the measurement eigenstates and we have assumed that $\langle\psi_+|\Psi\rangle = \langle\psi_-|\Psi\rangle = 1/\sqrt{2}$ (the results are equally probable).

We want to know how $|\Phi_0\rangle$ evolves in time during the act of measurement. From our discussion in Section 2.6, we know that the application of the time evolution operator $\hat{U}$ to $|\Phi_0\rangle$ allows us to calculate the state vector at some later time, which we denote as $|\Phi\rangle$, according to the simple expression $|\Phi\rangle = \hat{U}|\Phi_0\rangle$, or

$$|\Phi\rangle = \frac{1}{\sqrt{2}}\left[\hat{U}|\psi_+\rangle|\phi_0\rangle + \hat{U}|\psi_-\rangle|\phi_0\rangle\right]. \tag{3.3}$$

We now have to figure out what the effect of $\hat{U}$ will be.

It is clear that if the instrument interacts with a quantum system which is already present in one of the measurement eigenstates ($|\psi_+\rangle$, say), then the total system (quantum system plus instrument) must evolve into a product quantum state given by $|\psi_+\rangle|\phi_+\rangle$. This is equivalent to saying that this interaction will always produce a $+$ result (the pointer always moves to the left). In this case, the effect of $\hat{U}$ on the initial product quantum state $|\psi_+\rangle|\phi_0\rangle$ *must* be to yield the result $|\psi_+\rangle|\phi_+\rangle$, i.e.

$$\hat{U}|\psi_+\rangle|\phi_0\rangle = |\psi_+\rangle|\phi_+\rangle. \tag{3.4}$$

Similarly,

$$\hat{U}|\psi_-\rangle|\phi_0\rangle = |\psi_-\rangle|\phi_-\rangle \tag{3.5}$$

Substituting these last two expressions into eqn (3.3) gives

$$|\Phi\rangle = \frac{1}{\sqrt{2}}\left(|\psi_+\rangle|\phi_+\rangle + |\psi_-\rangle|\phi_-\rangle\right). \tag{3.6}$$

We now seem to be no further forward than before the measurement was made. Equation (3.6) suggests that the measuring instrument evolves into a superposition state in which the pointer simultaneously points both to the left and the right. Collapsing the wavefunction of the system-plus-measuring-device would seem to require a further measurement. But then the whole argument can be repreated *ad infinitum*. Are we therefore locked into an endless chain of measuring processes? At what point does the chain stop (at what point does the wavefunction collapse)?

This problem is created by our inability to obtain a collapse of the wavefunction using the continuous, deterministic equation of motion from which the time evolution operator $\hat{U}$ is derived (see Section 2.6). Schrödinger called the state vector $|\Phi\rangle$ as given in eqn (3.6) 'entangled' because, once generated, it is impossible to separate it into its constituent parts except by invoking an indeterministic collapse. As we have seen, such a collapse is simply not accounted for in the equations of orthodox quantum theory.

The paradox of Schrödinger's cat was designed to show up the apparent absurdity of this situation by shifting the focus from the microscopic world of sub-atomic particles to the macroscopic world of cats and human observers. The essential ingredients are shown in Fig. 3.5. A cat is placed inside a steel chamber together with a Geiger tube containing a small amount of radioactive substance, a hammer mounted on a pivot and a phial of prussic acid. The chamber is closed. From the amount of radioactive substance used and its known half-life, we expect that within one hour there is a probability of $\frac{1}{2}$ that one atom has disintegrated. If an atom does indeed disintegrate, the Geiger counter is triggered, releasing the hammer which smashes the phial. The prussic acid is released, killing the cat.

Prior to actually measuring the disintegration, the state vector of the atom or radioactive substance must be expressed as a linear superposition of the measurement eigenstates, corresponding to the physical states of the intact atom and the disintegrated atom. However, as we have seen above, treating the measuring instrument as a quantum object and using

Fig. 3.5 Schrödinger's cat.

the equations of quantum mechanics leads us to eqn (3.6), a superposition of the two possible outcomes of the measurement.

But what about the cat? These arguments would seem to suggest that we should express the state vector of the system-plus-cat as a linear superposition of the products of the state vectors describing a disintegrated atom and a dead cat and of the state vectors describing an intact atom and a live cat. In fact, the state vector of the dead cat is in turn a shorthand for the state corresponding to the triggered Geiger counter, released hammer, smashed phial, released prussic acid and dead cat. Prior to measurement, the physical state of the cat is therefore 'blurred' — it is neither alive nor dead but some peculiar combination of both states. We can perform a measurement on the cat by opening the chamber and ascertaining its physical state. Do we suppose that, at that point, the state vector of the system-plus-cat collapses and we record the observation that the cat is alive or dead as appropriate?

Although obviously intended to be somewhat tongue-in-cheek, Schrödinger's paradox nevertheless brings our attention to an important difficulty that we must confront. The Copenhagen interpretation says that elements of an empirical reality are defined by the nature of the experimental apparatus we construct to perform measurements on a quantum system. It insists that we resist the temptation to ask what physical state a particle (or a cat) was actually in prior to measurement as such a question is quite without meaning.

However, this positivist interpretation sits uncomfortably with some scientists, particularly those with a special fondness for cats. Some have accepted the EPR argument that quantum theory is incomplete. They have set about searching for an alternative theory, one that allows us to attach physical significance to the properties of particles without the need to specify the nature of the measuring instrument, one that allows us to define an independent reality and that reintroduces strict causality. Even though searching for such a theory might be engaging in meaningless metaphysical speculation, they believe that it is a search that has to be undertaken.

3.5 HIDDEN VARIABLES

If we reject the 'spooky' action at a distance that seems to be required in the Copenhagen interpretation of quantum theory, and which is highlighted by the EPR thought experiment, then we must accept the EPR argument that the theory is somehow incomplete. In essence, this involves the rejection of the first postulate of quantum theory: the state of a quantum mechanical system is *not* completely described by the wavefunction.

Those physicists who in the 1930s were uncomfortable with the Copenhagen interpretation were faced with two options. Either they could scrap quantum theory completely and start all over again or they could try to extend the theory to reintroduce strict causality and local reality. There was a general recognition that quantum theory was too good to be consigned to history's waste bin of scientific ideas. The theory did an excellent job of rationalizing the available experimental information on the physics of the microscopic world of quantum particles, and its predictions had been shown to be consistently correct. What was needed, therefore, was some means of adapting the theory to bring back those aspects of classical physics that it appeared to lack.

Einstein had hinted at a statistical interpretation. In his opinion, the squares of the wavefunctions of quantum theory represented statistical probabilities obtained by averaging over a large number of real particles. The obvious analogy here is with Boltzmann's statistical mechanics, which allows the calculation of observable physical quantities (such as gas pressure and thermodynamic functions like entropy) using atomic or molecular statistics. Although the theory deals with probabilities, these are derived from the behaviour of an ensemble of atoms or molecules which individually exist in predetermined physical states and which obey the laws of a deterministic classical mechanics.

The Copenhagen interpretation of the EPR experiment insists that the reality of the physical states that can be measured is defined by the nature of the interaction between two quantum particles and the nature of the experimental arrangement. A completely deterministic, locally real version of quantum theory demands that the physical states of the particles are 'set' at the moment of their interaction, and that the particles separate as individually real entities in those physical states. The physical states of the particles are fixed and independent of how we choose to set up the measuring instrument, and so no reference to the nature of the latter is necessary except to define how the independently real particles interact with it. The instrument thus *probes* an observer-independent reality.

Quantum theory in the form taught to undergraduate students of chemistry and physics tells us nothing about such physical states. This is either because they have no basis in reality (Copenhagen interpretation) or because the theory is incomplete (EPR argument). One way in which quantum theory can be made 'complete' in this sense is to introduce a new set of variables. These variables determine which physical states will be preferred as a result of a quantum process (such as an emission of a photon or a collision between two quantum particles). As these variables are not revealed in laboratory experiments, they are necessarily 'hidden' from us.

Hidden variable theories of one form or another are not without precedent in the history of science. Any theory which rationalizes the behaviour of a system in terms of parameters that are for some reason inaccessible to experiment is a hidden variable theory. These variables have often later become 'unhidden' through the application of new experimental technologies. The obvious example is again Boltzmann's use of the 'hidden' motions of real atoms and molecules to construct a statistical theory of mechanics. Mach's opposition to Boltmann's ideas was based on the extreme view that introducing such hidden variables unnecessarily complicates a theory and takes science no further forward. History has shown Mach's views to have been untenable.

We should note that although the introduction of hidden variables in quantum theory appears to be consistent with Einstein's general outlook, it has been claimed that Einstein himself never advocated such an approach. He appears to have been convinced that solutions to the conceptual problems of quantum theory would be found in an elusive grand unified field theory, the search for which took up most of his intellectual energy in the last decades of his life. However, if we exclude hidden variables, it is very difficult to imagine just what EPR must have had in mind when they argued that quantum theory is incomplete.

Statistical probabilities

It will help our discussion of hidden variables to run through a simple example in which we deduce and use statistical probabilities from an 'everyday' classical perspective, and then see how these compare with their equivalents in quantum theory. Imagine that I toss a coin into the air and it falls to the ground. When the coin comes to rest flat on the ground I take note of the outcome — heads (+ result, R_+) or tails (− result, R_-) — and enter this in my laboratory notebook. I repeat this 'measurement' process N times where N is a large number.

At the end of this experiment, I add up the total number of times the result R_+ was obtained and denote this as N_+. Similarly, N_- denotes the number of times R_- was obtained. As there are only two possible outcomes for each measurement, I know that $N_+ + N_- = N$. We define the frequencies of the results R_+ and R_- according to the relations:

$$\nu_+ = \frac{N_+}{N_+ + N_-}, \quad \nu_- = \frac{N_-}{N_+ + N_-}. \tag{3.7}$$

In principle, the outcome of any one measurement is determined by a number of variables, including the force and torque exerted on the coin as I toss it into the air, the interactions between the spinning coin and fluctuations in air currents and the angle and force of impact as the coin hits the ground. These variables could be controlled, for example by using a computer operated mechanical hand to toss the coin and by performing measurements in a vacuum. Alternatively, if we knew these variables precisely we could, in principle, use this information to calculate the exact trajectory of the coin. It is certainly not impossible that the outcome of a particular measurement could therefore be predicted with certainty.

In the absence of such control or knowledge of the variables, we assume that our measurements on the coin serve to 'calibrate' the system and allow us to make predictions about its behaviour in experiments yet to be performed. For example, if we discover that $\nu_+ = \nu_- = \frac{1}{2}$, we would conclude that the probabilities, P_+ and P_-, of obtaining the results R_+ and R_- respectively for the $(N+1)$th measurement would also be equal to $\frac{1}{2}$. We would further conclude that the coin is 'neutral' with regard to the measurement; i.e. both possible outcomes are obtained with equal probability. The coin does not have to be neutral: it could have been loaded in favour of one of the results and this would have been reflected in the measured frequencies. We should note that the definition of probability that we are using here is a rather intuitive one. In practice, coin tossing is subject to chance fluctuations that can often lead to some completely unexpected sequences of results. However, for our present purposes, it is sufficient to propose that our coin and method of tossing are unbiased and that we can make N sufficiently large so that the effects of chance fluctuations are averaged out.

The expectation value for the measurement is given by

$$\langle M_\pm \rangle = \frac{P_+ R_+ + P_- R_-}{P_+ + P_-}. \qquad (3.8)$$

Having established that $P_+ = P_- = \frac{1}{2}$, we conclude that the expectation value for the next measurement (and indeed all future measurements) is

$$\langle M_\pm \rangle = \frac{1}{2}(R_+ + R_-), \qquad (3.9)$$

i.e. we expect to obtain the result R_+ or the result R_- with equal probability.

We should make one further comment on eqn (3.9) before going on to consider a quantum system. Even when we do not control the variables as in this example, we perhaps have no difficulty in accepting that the outcome of a particular measurement is predetermined the moment the coin is launched into the air. Just as for Boltzmann's statistical mechanics, it is our lack of knowledge of the many individual variables at work which forces us to resort to statistical probabilities.

Quantum probabilities

Now consider a quantum system, described by the state vector $|\Psi\rangle$, on which a measurement also has two possible outcomes. Examples of such a measurement are the determination of the direction of an electron spin vector in some arbitrary laboratory frame and the determination of vertical versus horizontal polarization components of a circularly polarized photon. Our measuring instrument – operator $\hat{M}$ – has eigenstates $|\psi_+\rangle$ and $|\psi_-\rangle$ corresponding to the two possible outcomes: $\hat{M}|\psi_+\rangle = R_+|\psi_+\rangle$ and $\hat{M}|\psi_-\rangle = R_-|\psi_-\rangle$. To calculate the expectation value $\langle M_{\pm} \rangle$ we must first express the state vector $|\Psi\rangle$ in terms of the two measurement eigenstates using the expansion theorem:

$$|\Psi\rangle = |\psi_+\rangle \langle \psi_+|\Psi\rangle + |\psi_-\rangle \langle \psi_-|\Psi\rangle. \qquad (3.10)$$

It follows that

$$\hat{M}|\Psi\rangle = \hat{M}|\psi_+\rangle \langle \psi_+|\Psi\rangle + \hat{M}|\psi_-\rangle \langle \psi_-|\Psi\rangle \qquad (3.11)$$

$$= R_+|\psi_+\rangle \langle \psi_+|\Psi\rangle + R_-|\psi_-\rangle \langle \psi_-|\Psi\rangle$$

and so

$$\langle \Psi|\hat{M}|\Psi\rangle = R_+\langle \Psi|\psi_+\rangle \langle \psi_+|\Psi\rangle + R_-\langle \Psi|\psi_-\rangle \langle \psi_-|\Psi\rangle$$

$$= R_+\langle \psi_+|\Psi\rangle^*\langle \psi_+|\Psi\rangle + R_-\langle \psi_-|\Psi\rangle^*\langle \psi_-|\Psi\rangle$$

$$= |\langle \psi_+|\Psi\rangle|^2 R_+ + |\langle \psi_-|\Psi\rangle|^2 R_-$$

$$= P_+R_+ + P_-R_- \qquad (3.12)$$

where $P_+ = |\langle \psi_+|\Psi\rangle|^2$ is the projection probability for the eigenstate $|\psi_+\rangle$ and $P_- = |\langle \psi_-|\Psi\rangle|^2$ is the projection probability for the eigenstate $|\psi_-\rangle$. It can similarly be shown that $\langle \Psi|\Psi\rangle = P_+ + P_-$, and so

$$\langle M_{\pm} \rangle = \frac{\langle \Psi|\hat{M}|\Psi\rangle}{\langle \Psi|\Psi\rangle} = \frac{P_+R_+ + P_-R_-}{P_+ + P_-}. \qquad (3.13)$$

This is identical with the result obtained in eqn (3.9) and, in fact, reinforces the point made in Chapter 2 that the expression for the quantum theoretical expectation value is derived from an equivalent expression in probability calculus.

Equation (3.13) differs from eqn (3.9) only in the interpretation of the probabilities P_+ and P_-. In quantum theory, these quantities reflect the probabilities that the state vector $|\Psi\rangle$ collapses into one of the measurement eigenstates. Note that nowhere in the quantum theoretical analysis is it necessary to consider the behaviour of more than one quantum particle: eqn (3.13) applies to all individual particles in the state $|\Psi\rangle$.

A simple example of hidden variables

A photon in a state of left circular polarization is described by the state vector $|\psi_L\rangle$. We know that its interaction with a linear polarization analyser, and its subsequent detection, will reveal the photon to be in a state of vertical or horizontal polarization. Suppose, then, that the photon is completely described by $|\psi_L\rangle$ supplemented by some hidden variable λ which predetermines which state of linear polarization will be observed experimentally. By definition, λ itself is inaccessible to us through experiment, but its value somehow controls the way in which the photon interacts with the analyser.

We could imagine that λ has all the properties we would normally associate with a linear polarization vector. λ (or its projection) could presumably take up any angle in the plane perpendicular to the direction of propagation, as shown in Fig. 3.6(a). In a large ensemble of N left-circularly polarized photons, there would be a distribution of λ values over the N photons spanning the full 360° range. Thus, photon 1 has a λ value which we characterize in terms of the angle φ_1 it makes with the vertical axis, photon 2 has a λ value characterized by φ_2, and so on until we reach photon N, which has a λ value characterized by φ_N. These angles lie in the range 0°–360°.

We now need further to suppose that these λ values control the passage of the photons through the polarization analyser. A simple mechanism is as follows. If the angle that λ makes with the vertical axis lies within ±45° of that axis, then the photon passes through the vertical channel of the analyser and is detected as a vertically polarized photon (Fig. 3.6(b)). If, however, λ makes an angle with the vertical axis which lies outside this range, then the photon passes through the horizontal channel of the analyser and is detected as a horizontally polarized photon (Fig. 3.6(c)). We would need to suggest that the photon retains some 'memory' of its original circular polarization if we are to avoid the kinds of problems described in Section 2.6, which arise when two calcite crystals are placed 'back-to-back'. In fact, why not suppose that when the linear polarization properties of the photon become revealed, its circular polarization properties become hidden, controlled by another hidden variable.

In this scheme, the probability of detecting a photon in a state of vertical polarization becomes equal to the probability that the photon has a λ value within ±45° of the vertical axis. If there is a uniform probability that the λ value lies in the range 0°–360°, then the probability that it lies within ±45° of the vertical axis is clearly $\frac{1}{2}$. Similarly, the probability of detecting a photon in a state of horizontal polarization is also $\frac{1}{2}$. Thus, this simple hidden variable theory predicts results consistent with those of quantum theory.

Note that while we are still referring here to probabilities, unlike the probabilities of quantum theory these are now statistical, averaged over a large number of photons which individually possess clearly defined and predetermined properties. If the hidden variable approach were proved to be correct, we would presumably be able to trace these probabilities back to the (deterministic) physics of the processes that created the photons.

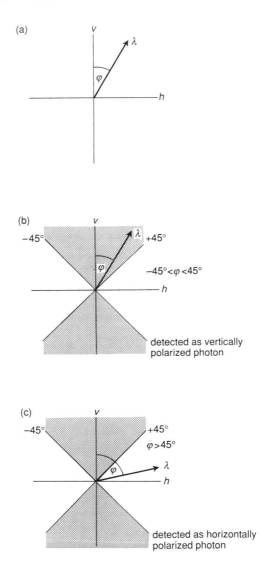

Fig. 3.6   (a) A simple example of a local hidden variable. The hidden vector λ determines the behaviour of a circularly polarized photon when it interacts with a linear polarization analyser. (b) If λ lies within $\pm45°$ of the vertical axis of the analyser, it passes through the vertical channel. (c) If it lies within $\pm45°$ of the horizontal axis, it passes through the horizontal channel.

We should not get too carried away with this simple scheme. While it does produce results consistent with the predictions of quantum theory, it will not explain some fairly rudimentary experimental facts of life, such as Malus's law. However, we expect that a little ingenuity on the part of theoretical physicists should soon get around this difficulty, albeit at the cost of introducing further complexity into the hidden variable theory.

Our simple example is obviously rather contrived and we would, perhaps, be reluctant at this stage to attach any physical significance to the variable λ, which can have whatever properties we like provided the end results agree with experiment. Nevertheless, this exercise at least seems to indicate that some kind of hidden variable scheme is feasible. It might therefore come as something of a shock to discover that John von Neumann demonstrated long ago that all such hidden variables are 'impossible'.

Von Neumann's 'impossibility proof'

In a hidden variable extension of quantum theory, an ensemble of N quantum particles, all described by some state vector $|\Psi\rangle$, contains particles with some distribution of λ values. For an individual particle, the value of λ predetermines its behaviour during the measurement process. Let us suppose that the result of a measurement (operator $\hat{M}$) is one of two possibilities, R_+ and R_-. The ensemble N can then be divided into two sub-ensembles, which we denote N_+ and N_-. The sub-ensemble N_+ consists of those particles with λ values which predetermine the result R_+ for each particle. The sub-ensemble N_- similarly contains only those particles predisposed to give the result R_-. Referring to our simple example given above, N_+ would contain all those photons with λ values characteristic of vertical polarization, and N_- would contain those photons with λ values characteristic of horizontal polarization.

If we perform measurements only on the sub-ensemble N_+, we know that we should always obtain the result R_+. Such an ensemble is said to be dispersion free. A dispersion-free ensemble has the property that

$$\langle M_\pm^2 \rangle - \langle M_\pm \rangle^2 = 0 \qquad (3.14)$$

where $\langle M_\pm \rangle$ is the expectation value for the result obtained by operating on the state vector $|\Psi\rangle$ with $\hat{M}$. Von Neumann's proof rests on the demonstration that such dispersion-free ensembles are impossible, and hence no hidden variable theory can reproduce the results that are so readily explained by quantum theory.

We should first confirm that eqn (3.14) is not true for $|\Psi\rangle$ when expressed as a linear superposition of the eigenstates of the measurement operator, as required in orthodox quantum theory without hidden variables. We have

$$\hat{M}^2 | \Psi \rangle = \hat{M}^2 | \psi_+ \rangle \langle \psi_+ | \Psi \rangle + \hat{M}^2 | \psi_- \rangle \langle \psi_- | \Psi \rangle$$

$$= R_+^2 | \psi_+ \rangle \langle \psi_+ | \Psi \rangle + R_-^2 | \psi_- \rangle \langle \psi_- | \Psi \rangle \quad (3.15)$$

and so

$$\langle M_\pm^2 \rangle = \langle \Psi | \hat{M}^2 | \Psi \rangle = P_+ R_+^2 + P_- R_-^2 \quad (3.16)$$

where P_+ and P_- were defined above. Obviously, from eqn (3.13), $\langle M_\pm^2 \rangle \neq \langle M_\pm \rangle^2$ (remember $\langle \Psi | \Psi \rangle = P_+ + P_- = 1$) and the ensemble exhibits dispersion.

If the hidden variables are to have the intended effect, the expectation value of the measurement operator *must* be equal to one of its eigenvalues. This follows automatically from our requirement that a quantum particle be described by $| \Psi \rangle$ and some hidden variable which predetermines the result R_+ or R_-. This requirement means that for a particle in the sub-ensemble N_+, the effect of $\hat{M}$ on $| \Psi \rangle$ must be to return *only* the eigenvalue R_+. Thus,

$$\hat{M} | \Psi \rangle_{N_+} = R_+ | \Psi \rangle_{N_+} \quad (3.17)$$

where we have used a subscript N_+ to indicate that this expression applies only to the sub-ensemble N_+. From eqn (3.17) it follows that

$$\langle M_\pm \rangle_{N_+} = \langle \Psi | \hat{M} | \Psi \rangle_{N_+} = R_+. \quad (3.18)$$

Similarly,

$$\langle M_\pm^2 \rangle_{N_+} = \langle \Psi | \hat{M}^2 | \Psi \rangle_{N_+} = R_+^2. \quad (3.19)$$

and so $\langle M_\pm^2 \rangle_{N_+} = \langle M_\pm \rangle_{N_+}^2$ and the sub-ensemble is dispersion free, as required.

Von Neumann's mathematical proof is quite complicated and we will deal with it here only in a superficial manner. Interested readers are advised to consult the more advanced texts given in the bibliography. The proof is based on a number of postulates, one of which merits our attention. Imagine that an operator $\hat{O}$ corresponding to some physical quantity can be written as a combination of other operators (for example, $\hat{H} = \hat{T} + V$). Von Neumann postulated that the expectation value $\langle O \rangle$ can be obtained as a linear combination of the expectation values of the operators that combine to make up $\hat{O}$, whether or not these operators commute. Thus, in general, for $O = aA + bB + \dots$

$$\langle O \rangle = \langle aA + bB + \dots \rangle = a \langle A \rangle + b \langle B \rangle + \dots \quad (3.20)$$

It is relatively straightforward to show that this is indeed the case for operators in quantum theory.

Von Neumann then considered the measurement of a second, complementary physical quantity (operator $\hat{L}$) on the sub-ensemble N_+. We suppose that there are again two possible outcomes, results S_+ and S_-. Following the same line of argument, we need to propose that there are two sub-sub-ensembles of N_+, one of which is predisposed to give only the result S_+ and one which gives only S_-. We denote these two sub-sub-ensembles as N_{++} and N_{+-}. Using eqn (3.20), we can write

$$\langle M_\pm + L_\pm \rangle_{N_{++}} = \langle M_\pm \rangle_{N_{++}} + \langle L_\pm \rangle_{N_{++}}$$
$$= R_+ + S_+. \tag{3.21}$$

Herein lies the difficulty, von Neumann claimed. Note that, unlike the equation

$$\langle M_\pm \rangle = \frac{1}{2}(R_+ + R_-), \tag{3.9}$$

the expectation value of the combined operator $\hat{M} + \hat{L}$ is given by the sum of two eigenvalues corresponding to two measurement processes each of which must be obtained with unit probability (i.e. with certainty). Whereas eqn (3.9) is interpreted to mean that the result R_+ *or* the result R_- may be obtained with equal probability, eqn (3.21) can *only* mean that R_+ *and* S_+ *must* each be obtained with unit probability. However, although the expectation values of non-commuting quantum mechanical operators are additive, as postulated in eqn (3.20), their eigenvalues are not. If they were, then an appropriate choice of measurement operators would allow us simultaneously to measure the position and momentum of a quantum particle with arbitrary precision, or mutually exclusive electron spin orientations, or simultaneous linear and circular polarization states of photons. This conflicts with experiment. That the expectation values $\langle M_\pm \rangle_{N_{++}}$ and $\langle L_\pm \rangle_{N_{++}}$ are equal to the eigenvalues of the corresponding operators is a requirement if the sub-sub-ensemble N_{++} is to be dispersion free. Von Neumann therefore concluded that dispersion-free ensembles (and hence hidden variables) are impossible.

Von Neumann was congratulated not only by his colleagues and those fellow physicists who favoured the Copenhagen interpretation, but also by his opponents. However, if this were the end of the story as far as hidden variable theories are concerned, then we could eliminate virtually all of Chapter 4 from this book. Von Neumann's impossibility proof certainly discouraged the physics community from taking the idea of hidden variables seriously, although a few (notably Schrödinger and de Broglie) were not put off by it. Others began to look closely at the proof and became suspicious. A few questioned the proof's correctness. The physicist Grete Hermann suggested that von Neumann's proof was circular—that it presupposed what it was trying to prove in its premises. She argued that the additivity postulate, eqn (3.20), while certainly true for quantum states in ordinary quantum theory, cannot be automatically assumed to hold for states described in terms of hidden variables. Since von Neumann's proof rests on the general non-additivity of eigenvalues, it collapses without the additivity postulate.

In his book *The philosophy of quantum mechanics*, published in 1974, Max Jammer examined Hermann's arguments and concluded that the charge of circularity is not justified. He noted that the additivity postulate was intended to apply to all operators, not just non-commuting operators (which would give rise to non-additive eigenvalues). However, for commuting operators the case against dispersion-free states is not proven by von Neumann's arguments. Jammer wrote:[†] 'What should have been criticised, instead, is the fact that the proof severely restricts the class of conceivable ensembles by admitting only those for which [the additivity postulate] is valid.'

It is also worth noting an objection raised by the physicist John S. Bell (who we will meet again in the next chapter). Bell argued that von Neumann's proof applies to the simultaneous measurement of two complementary physical quantities. But such measurements require completely incompatible measuring devices and so no-one should be surprised if the corresponding eigenvalues are not additive.

It gradually began to dawn on the physics community that hidden variables were not impossible after all. But about 20 years passed between the publication of von Neumann's proof and the resurgence of interest in hidden variable theories. By that time the Copenhagen interpretation was well entrenched in quantum physics and those arguing against it were in a minority.

[†] Jammer, Max (1974). *The philosophy of quantum mechanics.* John Wiley and Sons, NY.

4
Putting it to the test

4.1 BOHM'S VERSION OF THE EPR EXPERIMENT

Work on hidden variable solutions to the conceptual problems of quantum theory did not exactly stop after the publication of von Neumann's 'impossibility proof', but then it hardly represented an expanding field of scientific activity. About 20 years elapsed before David Bohm, a young American physicist, began to take more than a passing interest in the subject. His first, all-important contributions to the debate over the interpretation of quantum theory were made in 1951.

In February of that year he published a book, simply entitled *Quantum theory*, in which he presented a discussion of the EPR thought experiment. At that stage, he appeared to accept Bohr's response to EPR as having settled the matter in favour of the Copenhagen interpretation. He wrote:[†] 'Their [EPR's] criticism has, in fact, been shown to be unjustified, and based on assumptions concerning the nature of matter which implicitly contradict the quantum theory at the outset.' But the subtle nature of the EPR argument, and the apparently natural and common-sense assumptions behind it, encouraged Bohm to analyse the argument in some detail. In this analysis, he made extensive use of a derivative of the EPR thought experiment that ultimately led other physicists to believe that it could be brought down from the lofty heights of pure thought and put into the practical world of the physics laboratory. It is this aspect of Bohm's contribution that we will consider here.

Bohm's work on the EPR argument set him thinking deeply about the problems of the Copenhagen interpretation. He was very soon tinkering with hidden variables, and his first papers on this subject were submitted to *Physical Review* in July 1951, only four months after the publication of his book. However, Bohm's hidden variables differ from the ones we have so far considered (and with which we will stay in this chapter) in that they are non-local. We examine Bohm's non-local hidden variable theory in Chapter 5.

[†] Bohm, David (1951). *Quantum theory*. Prentice-Hall, Englewood Cliffs, NJ.

Correlated spins

Bohm considered a molecule consisting of two atoms in a quantum state in which the total electron spin angular momentum is zero. A simple example would be a hydrogen molecule with its two electrons spin-paired in the lowest (ground) electronic state (see Fig. 4.1). We suppose that we can dissociate this molecule in a process that does not change the total angular momentum to produce two equivalent atomic fragments. The hydrogen molecule splits into two hydrogen atoms. These atoms move apart but, because they are produced by the dissociation of an excited molecule with no net spin and, by definition, the spin does not change, the spin orientations of the electrons in the individual atoms remain opposed.

The spins of the atoms themselves are therefore correlated. Measurement of the spin of one atom (say atom A) in some arbitrary laboratory frame allows us to predict, with certainty, the direction of the spin of atom B in the same frame. Viewed in terms of classical physics or via the perspective of local hidden variables, we would conclude that the spins of the two atoms are determined by the nature of the initial molecular quantum state and the method of dissociation. The atoms move away from each other with their spins fixed in unknown but opposite orientations and the measurement merely tells us what these orientations are.

In contrast, the two atoms are described in quantum theory by a single wavefunction or state vector until the moment of measurement. If we choose to measure the component of the spin of atom A along the laboratory z axis, our observation that the wavefunction is projected into a state in which atom A has its angular momentum vector aligned in the $+z$ direction (say) means that atom B must have its angular momentum vector aligned in the $-z$ direction. But what if we choose, instead, to measure the x or y components of the spin of atom A? No matter which component is measured, the physics of the dissociation demand that the

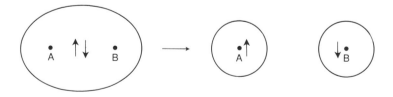

Fig. 4.1   Correlated quantum particles. The dissociation of a molecule from its ground state with no change of electron spin orientation creates a pair of atoms whose spins are correlated.

spins of the atoms must still be correlated, and so the opposite results must always be obtained for atom B. If we accept the definition of physical reality offered by EPR, then we must conclude that *all* components of the spin of atom B are elements of reality, since it appears that we can predict them with certainty without in any way disturbing B.

However, the wavefunction specifies only one spin component, associated with the magnetic spin quantum number m_s. This is because the operators corresponding to the three components of the spin vector in Cartesian coordinates do not commute (the components are complementary observables). Thus, either the wavefunction is incomplete, or EPR's definition of physical reality is unjustified. The Copenhagen interpretation says that no spin component of atom B 'exists' until a measurement is made on atom A. The result we obtain for B will depend on how we choose to set up our instrument to make measurements on A. This is entirely consistent with EPR's original argument, couched in terms of the complementary position–momentum observables of two correlated particles. However, the measurement of the spin component of an atom (or an electron) is much more practicable than the measurement of the position or momentum of an atom. Some physicists saw that further elaborations of Bohm's version of the EPR experiment could be carried out in the laboratory. We examine one of these next.

Correlated photons

It is convenient to extend Bohm's version of the EPR experiment further. Suppose an atom in an electronically excited state emits two photons in rapid succession as it returns to the ground state. Suppose also that the total electron orbital and spin angular momentum of the atom in the excited state is the same as that in the ground state. Conservation of angular momentum demands that the net angular momentum carried away by the photons is zero.

We know from our discussion in Section 2.5 that all photons possess a spin quantum number $s = 1$ and can have 'magnetic' spin quantum numbers $m_s = \pm 1$, corresponding to states of left and right circular polarization. The net angular momentum of the photon pair can be zero only if the photons are emitted with opposite values of m_s, i.e. in opposite states of circular polarization. This scheme is exactly analogous to Bohm's version of the EPR experiment, but we have replaced the creation of a pair of atoms with opposite spin orientations with the creation of a pair of photons with opposite spin orientations (circular polarizations). We discuss how this can be achieved in practice in Section 4.4.

The experimental arrangement drawn in Fig. 4.2 is designed not to measure the circular polarizations of the photons but, instead, measures their vertical and horizontal polarizations. A photon moving to the left (photon A) passes through polarization analyser 1 (denoted PA_1). This analyser is oriented vertically (orientation a) with respect to some arbitrary laboratory frame. For reasons which will become clear when we go through a mathematical analysis below, the detection of photon A in a state of vertical polarization means that when B passes through polarization analyser 2 (PA_2, which also has orientation a), it *must* be measured also in a state of vertical polarization. This polarization state of B will be 180° out of phase with the corresponding state of A, because the net angular momentum of the pair must be zero, but such phase information is not recovered from the measurements. Similarly, the measurement of A in a state of horizontal polarization implies that B must be measured also in a state of horizontal polarization. We can therefore predict, with certainty, the vertical versus horizontal polarization state of B from measurements we make on A.

According to the Copenhagen interpretation, we know only the probabilities that an individual photon will be detected in a vertical or horizontal polarization state; its polarization direction is not predetermined by any property that the photon possesses prior to measurement. In contrast, according to any local hidden variable theory, the behaviour of each photon is governed by a hidden variable which precisely defines its polarization direction (along any axis) and the photon follows a predetermined path through the apparatus.

Mathematical analysis

To anticipate or interpret the results of such an experiment in terms of quantum theory, we need to know the initial state vector of the photon pair and the possible measurement eigenstates. We will begin with the former.

The two photons are emitted with opposite spin orientations or circular polarizations. The total state vector of the pair can therefore be written as a linear superposition of the product of the states $| \psi_L^A \rangle$ (photon A in a state of left circular polarization) and $| \psi_L^B \rangle$ (photon B in a state of left circular polarization) and the product of $| \psi_R^A \rangle$ (photon A in a state of right circular polarization) and $| \psi_R^B \rangle$ (photon B in a state of right circular polarization). Readers might have expected that these products should have been $| \psi_L^A \rangle | \psi_R^B \rangle$ and $| \psi_R^A \rangle | \psi_L^B \rangle$ to get the correct left–right symmetry, but remember that the convention for circular polarization given in Section 2.5 specifies the direction of rotation for photons propagating *towards* the detector. Left (anticlockwise) rotation with respect to PA_1 corresponds to right (clockwise) rotation with respect to PA_2, and so we interchange the labels for photon B.

We now need to recall that photons are bosons and from Section 2.4 we note that bosons have two-particle state vectors that are symmetric to the exchange of the particles. The initial state vector of the pair is therefore given by:

$$| \Psi \rangle = \frac{1}{\sqrt{2}} (| \psi_L^A \rangle | \psi_L^B \rangle + | \psi_R^A \rangle | \psi_R^B \rangle) \qquad (4.1)$$

The arrangement drawn in Fig. 4.2 can produce any one of four possible outcomes for each successfully detected pair. If we denote detection of a photon in a state of vertical polarization as a $+$ result and detection in a state of horizontal polarization as a $-$ result, these four measurement possibilities are:

PA$_1$	PA$_2$	Measurement eigenstate
$+$	$+$	$| \psi_{++} \rangle$
$+$	$-$	$| \psi_{+-} \rangle$
$-$	$+$	$| \psi_{-+} \rangle$
$-$	$-$	$| \psi_{--} \rangle$

The joint measurement eigenstates are the products of the final state vectors of the individual photons. Denoting these final states as $| \psi_v^A \rangle$ (photon A detected in vertical polarization state with respect to orientation a), $| \psi_h^A \rangle$ (photon A detected in horizontal polarization state with respect to orientation a), $| \psi_v^B \rangle$ and $| \psi_h^B \rangle$, we have

$$| \psi_{++} \rangle = | \psi_v^A \rangle | \psi_v^B \rangle \qquad | \psi_{+-} \rangle = | \psi_v^A \rangle | \psi_h^B \rangle$$
$$| \psi_{-+} \rangle = | \psi_h^A \rangle | \psi_v^B \rangle \qquad | \psi_{--} \rangle = | \psi_h^A \rangle | \psi_h^B \rangle . \qquad (4.2)$$

Now we must do something about the fact that the initial state vector $| \Psi \rangle$ is given in eqn. (4.1) in a basis of circular polarization states

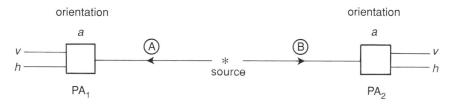

Fig. 4.2 Experimental arrangement to measure the polarization states of pairs of correlated photons.

whereas the measurement eigenstates are given in a basis of linear polarization states. We therefore use the expansion theorem to express the initial state vector in terms of the possible measurement eigenstates:

$$| \Psi \rangle = | \psi_{++} \rangle \langle \psi_{++} | \Psi \rangle + | \psi_{+-} \rangle \langle \psi_{+-} | \Psi \rangle + | \psi_{-+} \rangle \langle \psi_{-+} | \Psi \rangle$$
$$+ | \psi_{--} \rangle \langle \psi_{--} | \Psi \rangle . \tag{4.3}$$

We must now find expressions for the individual projection amplitudes in eqn (4.3). From eqns (4.1) and (4.2) we have

$$\langle \psi_{++} | \Psi \rangle = \frac{1}{\sqrt{2}} \langle \psi_v^A | \langle \psi_v^B | (| \psi_L^A \rangle | \psi_L^B \rangle + | \psi_R^A \rangle | \psi_R^B \rangle) \tag{4.4a}$$

$$= \frac{1}{\sqrt{2}} (\langle \psi_v^A | \psi_L^A \rangle \langle \psi_v^B | \psi_L^B \rangle + \langle \psi_v^A | \psi_R^A \rangle \langle \psi_v^B | \psi_R^B \rangle) \tag{4.4b}$$

$$= \frac{1}{\sqrt{2}} \left(\frac{1}{\sqrt{2}} \frac{1}{\sqrt{2}} + \frac{1}{\sqrt{2}} \frac{1}{\sqrt{2}} \right) \tag{4.4c}$$

$$= \frac{1}{\sqrt{2}} \tag{4.4d}$$

where we have used the information in Table 2.2 to obtain expressions for the circular–linear polarization state projection amplitudes that appear in eqn (4.4b). Repeating this process for the other projection amplitudes in eqn (4.3) gives

$$\langle \psi_{+-} | \Psi \rangle = 0$$
$$\langle \psi_{-+} | \Psi \rangle = 0 \tag{4.5}$$
$$\langle \psi_{--} | \Psi \rangle = - \frac{1}{\sqrt{2}}$$

and so

$$| \Psi \rangle = \frac{1}{\sqrt{2}} (| \psi_{++} \rangle - | \psi_{--} \rangle). \tag{4.6}$$

Equation (4.5) confirms our earlier view that the detection of a combined + result for photon A and − result for photon B (and vice versa) is not possible for the arrangement shown in Fig. 4.2 in which both polarization analysers have the same orientation. From eqn (4.6) we can deduce that the joint probability for both photons to produce + results, $P_{++}(a, a) = | \langle \psi_{++} | \Psi \rangle |^2$, is equal to the joint probability for both photons to give − results, $P_{--}(a, a) = | \langle \psi_{--} | \Psi \rangle |^2$, i.e. $P_{++}(a, a) = P_{--}(a, a) = \frac{1}{2}$. The notation (a, a) indicates the orientations of the two analysers.

The expectation value

We denote the measurement operator corresponding to PA_1 in orientation a as $\hat{M}_1(a)$. The results of the operation of $\hat{M}_1(a)$ on photon A are R_v^A or R_h^A, depending on whether A is detected in a final vertical or horizontal polarization state. Thus, we can write $\hat{M}_1(a)|\psi_v^A\rangle = R_v^A|\psi_v^A\rangle$ and $\hat{M}_1(a)|\psi_h^A\rangle = R_h^A|\psi_h^A\rangle$. Similarly, $\hat{M}_2(a)|\psi_v^B\rangle = R_v^B|\psi_v^B\rangle$ and $\hat{M}_2(a)|\psi_h^B\rangle = R_h^B|\psi_h^B\rangle$, where $\hat{M}_2(a)$ is the operator corresponding to PA_2 in orientation a and R_v^B and R_h^B are the corresponding eigenvalues. From eqn (4.6), we have

$$\hat{M}_2(a)|\Psi\rangle = \frac{1}{\sqrt{2}}\left(\hat{M}_2(a)|\psi_{++}\rangle - \hat{M}_2(a)|\psi_{--}\rangle\right)$$

$$= \frac{1}{\sqrt{2}}\left(R_v^B|\psi_{++}\rangle - R_h^B|\psi_{--}\rangle\right) \tag{4.7}$$

and so

$$\hat{M}_1(a)\hat{M}_2(a)|\Psi\rangle = \frac{1}{\sqrt{2}}\left(R_v^B\hat{M}_1(a)|\psi_{++}\rangle - R_h^B\hat{M}_1(a)|\psi_{--}\rangle\right)$$

$$= \frac{1}{\sqrt{2}}\left(R_v^A R_v^B|\psi_{++}\rangle - R_h^A R_h^B|\psi_{--}\rangle\right). \tag{4.8}$$

Thus, the expectation value for the joint measurement is given by

$$\langle\Psi|\hat{M}_1(a)\hat{M}_2(a)|\Psi\rangle = \frac{1}{2}\left(R_v^A R_v^B + R_h^A R_h^B\right). \tag{4.9}$$

This notation is getting rather cumbersome, so we will from now on abbreviate the expectation value in eqn (4.9) as $E(a,a)$. Note that the result in eqn (4.9) is equivalent to $E(a,a) = P_{++}(a,a)R_v^A R_v^B + P_{--}(a,a)R_h^A R_h^B$, where $P_{++}(a,a) = P_{--}(a,a) = \frac{1}{2}$. The *correlation* between the joint measurements is most readily seen if we ascribe some values to the individual results. For example, we can set $R_v^A = R_v^B = +1$ and $R_h^A = R_h^B = -1$, which is perfectly legitimate since we can always suppose that the measurement operators can be expressed in a way which reproduces these particular eigenvalues. Putting these results into eqn (4.9) gives

$$E(a,a) = +1, \tag{4.10}$$

i.e. the joint results are perfectly correlated.

A poor map of reality

If the discussion above has so far seemed reasonable, we must acknowledge one important point about it. Although there are some properties of $| \Psi \rangle$ that depend only on the nature of the physics of the two-photon emission and the atomic quantum states involved, our quantum theory analysis is useful to us only when couched in terms of the measurement eigenstates of the apparatus. There are no 'intrinsic' states of the quantum system. Even the initial state vector, given in eqn (4.1), is only meaningful if we relate it to some kind of experimental arrangement. Of course, quantum theory tells us nothing whatsoever about the 'real' polarization directions of the photons (these are properties that supposedly have no basis in reality). Consequently, the *only* way of treating $| \Psi \rangle$ is in relation to our measuring device.

For example, we could have aligned each polarization analyser to make measurements along one of many quite arbitrary directions. The arrangement shown in Fig. 4.2 measures the vertical v and horizontal h components of the photon polarizations. However, we could rotate both polarization analysers through any angle φ in the same direction and measure the v' and h' components. But, provided both analysers are aligned in the same direction, the observed results would be just the same. All polarization components are therefore possible, but only in an incompletely defined sense. To obtain a complete specification, the photons must interact with a device which defines the direction in which the components are to be measured and simultaneously excludes the measurement of all other components. Definiteness in one direction must lead to complete indefiniteness in all other directions (complementarity).

Bohm closed his discussion of his version of the EPR experiment with the comment:[†] 'Thus, we must give up the classical picture of a precisely defined [polarization] associated with each [photon], and replace it by our quantum concept of a potentiality, the probability of whose development is given by the wave function.' Bohm's reference to 'potentialities'—the potential inherent in a quantum system to produce a particular result—suggests that he may already have been thinking about non-local hidden variables, despite his outward adherence to the Copenhagen interpretation. He also noted that the mathematical formalism of quantum theory did not contain elements that provide a one-to-one correspondence with the actual behaviour of quantum particles. 'Instead', he wrote, 'we have come to the point of view that the wave function is an abstraction, providing a mathematical reflection of certain aspects of reality, but not a one-to-one mapping.'

He further concluded that: '. . . no theory of mechanically deter-

---

[†] Bohm, David (1951). *Quantum theory*. Prentice-Hall, Englewood Cliffs, NJ.

mined hidden variables can lead to *all* of the results of the quantum theory.'

4.2 QUANTUM THEORY AND LOCAL REALITY

The pattern of results observed for a beam of photons passing through a polarization analyser is not changed as we rotate the analyser. In principle, the measurement eigenstates $|\psi_v\rangle$ and $|\psi_h\rangle$ refer only to the direction 'imposed' on the quantum system by the apparatus itself — we need to use the notation v' and h' only when one analyser orientation differs from the other. The pattern does not depend on whether we orient the apparatus along the laboratory z axis, x axis or, indeed, any axis. However, important differences arise when two sets of apparatus are used to make measurements on correlated pairs of quantum particles, since the two sets of measurement eigenstates need not refer to the same direction.

Quantum correlations

Let us consider the effects of rotating PA_2 through some angle with respect to PA_1, as shown in Fig. 4.3. PA_1 is aligned in the same direc-

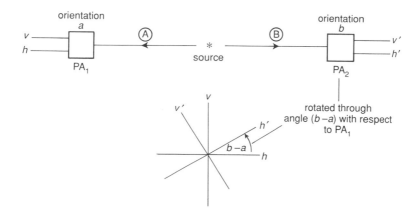

Fig. 4.3   The same arrangement as shown in Fig. 4.2, but with one of the polarization analysers oriented at an angle with respect to the vertical axis of the other.

tion as before, which we continue to denote as orientation a, and so its measurement eigenstates are $|\psi_v^A\rangle$ and $|\psi_h^A\rangle$, with eigenvalues R_v^A and R_h^A. We denote the orientation of PA_2 as b and designate its new measurement eigenstates as $|\psi_{v'}^B\rangle$ and $|\psi_{h'}^B\rangle$, corresponding respectively to polarization along the new vertical v' and horizontal h' directions, with corresponding eigenvalues $R_{v'}^A$, and $R_{h'}^B$. We denote the angle between the vertical axes of the analysers as $(b-a)$. The eigenstates of the joint measurement in this new arrangement are given by

$$|\psi'_{++}\rangle = |\psi_v^A\rangle |\psi_{v'}^B\rangle \qquad\qquad |\psi'_{+-}\rangle = |\psi_v^A\rangle |\psi_{h'}^B\rangle$$
$$|\psi'_{-+}\rangle = |\psi_h^A\rangle |\psi_{v'}^B\rangle \qquad\qquad |\psi'_{--}\rangle = |\psi_h^A\rangle |\psi_{h'}^B\rangle. \tag{4.11}$$

We must now express the initial state vector $|\Psi\rangle$, given in eqn (4.1), in terms of the the new joint measurement eigenstates. We can obviously proceed in the same way as before:

$$|\Psi\rangle = |\psi'_{++}\rangle\langle\psi'_{++}|\Psi\rangle + |\psi'_{+-}\rangle\langle\psi'_{+-}|\Psi\rangle + |\psi'_{-+}\rangle\langle\psi'_{-+}|\Psi\rangle$$
$$+ |\psi'_{--}\rangle\langle\psi'_{--}|\Psi\rangle \tag{4.12}$$

in which

$$\langle\psi'_{++}|\Psi\rangle = \frac{1}{\sqrt{2}}\langle\psi_v^A|\langle\psi_{v'}^B|(|\psi_L^A\rangle|\psi_L^B\rangle + |\psi_R^A\rangle|\psi_R^B\rangle) \tag{4.13a}$$

$$= \frac{1}{\sqrt{2}}(\langle\psi_v^A|\psi_L^A\rangle\langle\psi_{v'}^B|\psi_L^B\rangle + \langle\psi_v^A|\psi_R^A\rangle\langle\psi_{v'}^B|\psi_R^B\rangle) \tag{4.13b}$$

$$= \frac{1}{\sqrt{2}}\left(\frac{1}{\sqrt{2}}\frac{e^{-i(b-a)}}{\sqrt{2}} + \frac{1}{\sqrt{2}}\frac{e^{i(b-a)}}{\sqrt{2}}\right) \tag{4.13c}$$

$$= \frac{1}{\sqrt{2}}\cos(b-a). \tag{4.13d}$$

Similarly,

$$\langle\psi'_{+-}|\Psi\rangle = \frac{1}{\sqrt{2}}\sin(b-a)$$

$$\langle\psi'_{-+}|\Psi\rangle = \frac{1}{\sqrt{2}}\sin(b-a)$$

$$\langle\psi'_{--}|\Psi\rangle = \frac{1}{\sqrt{2}}\cos(b-a) \tag{4.14}$$

Thus,

$$|\Psi\rangle = \frac{1}{\sqrt{2}}\left[\,|\psi_{++}\rangle\cos(b-a) + |\psi_{+-}\rangle\sin(b-a)\right.$$

$$\left. + |\psi_{-+}\rangle\sin(b-a) - |\psi_{--}\rangle\cos(b-a)\,\right]. \tag{4.15}$$

The consistency of this last expression with the result we obtained in eqn (4.6) can be confirmed by setting $b = a$. We can use eqns (4.13d) and (4.14) to obtain the probabilities for each of the four possible joint results:

$$P_{++}(a,b) = |\langle\psi'_{++}|\Psi\rangle|^2 = \frac{1}{2}\cos^2(b-a)$$

$$P_{+-}(a,b) = |\langle\psi'_{+-}|\Psi\rangle|^2 = \frac{1}{2}\sin^2(b-a)$$

$$P_{-+}(a,b) = |\langle\psi'_{-+}|\Psi\rangle|^2 = \frac{1}{2}\sin^2(b-a)$$

$$P_{--}(a,b) = |\langle\psi'_{--}|\Psi\rangle|^2 = \frac{1}{2}\cos^2(b-a). \tag{4.16}$$

The expectation value, $E(a,b)$, is given by

$$E(a,b) = P_{++}(a,b)R_v^A R_v^B + P_{+-}(a,b)R_v^A R_{h'}^B + P_{-+}(a,b)R_h^A R_{v'}^B$$
$$+ P_{--}(a,b)R_h^A R_{h'}^B. \tag{4.17}$$

If, as before, we ascribe values of ± 1 to the individual results ($+1$ for v or v' polarization, -1 for h or h' polarization), then the expectation value can be used as a measure of the correlation between the joint measurements:

$$E(a,b) = P_{++}(a,b) - P_{+-}(a,b) - P_{-+}(a,b) + P_{--}(a,b). \tag{4.18}$$

From eqn (4.16), we discover that $E(a,b)$ for the experimental arrangement shown in Fig. 4.3 is given by

$$E(a,b) = \cos^2(b-a) - \sin^2(b-a) = \cos 2(b-a). \tag{4.19}$$

The function $\cos 2(b-a)$ is plotted against $(b-a)$ in Fig. 4.4. Note how this function varies between $+1$ ($b = a$, perfect correlation), through 0 ($(b-a) = 45°$, no correlation) to -1 ($(b-a) = 90°$, perfect anticorrelation).

Hidden variable correlations

What are the predictions for $E(a,b)$ using a local hidden variable theory? We will answer this question here by reference to the very simple local hidden variable theory described in Section 3.5. We suppose that

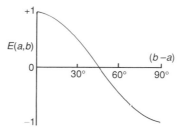

Fig. 4.4 The correlation between the photon polarization states predicted by quantum theory, plotted as a function of the angle between the vertical axes of the analysers.

the two photons are emitted with opposite circular polarizations, as required by the physics of the emission process, but that they also possess fixed values of some hidden variables which predetermine their linear polarization states.

As before, we imagine that these hidden variables behave rather like linear polarization vectors. Thus, after emission, photon A might move towards PA_1 in a quantum state which we could denote $|\psi_L^A, \lambda\rangle$, indicating that it is left circularly polarized and has a value of λ which predetermines its linear polarization state in one specific direction. This value of λ is set at the moment of emission and remains fixed as the photon moves towards PA_1. Consequently, photon B must move towards PA_2 in the state $|\psi_R^B, -\lambda\rangle$, indicating that it is right circularly polarized and has a value of λ which is opposite to that of A (no net angular momentum) but which predetermines that its linear polarization state lies in the same (vertical) plane as that of A. As with A, the λ value of B is set at the moment of emission and remains fixed as it moves towards PA_2. Its value is not changed on detection of photon A (Einstein separability).

According to our simple theory, a photon with λ pointing in any direction within $\pm 45°$ of the vertical axis of a polarizer will pass through the vertical channel. If it lies outside this range, then it must lie within $\pm 45°$ of the horizontal axis and so passes through the horizontal channel of the polarizer.

We set the two analysers so that they are aligned in the same direction $(b = a)$. For simplicity, we imagine the situation where photon A passes through the vertical channel of PA_1 (+ result). This means that the λ value of A must have been within $\pm 45°$ of the vertical axis. The λ value of photon B, which points in the opposite direction, must therefore lie within $\pm 45°$ of the vertical axis of PA_2 and so passes through the vertical channel of PA_2, as shown in Fig. 4.5. Hence the two photons produce a joint + + result, consistent with the quantum theory predic-

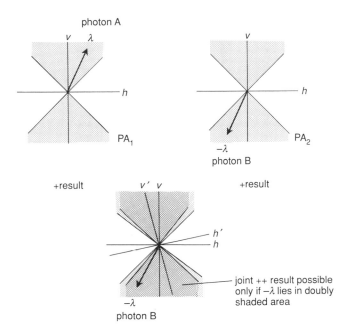

Fig. 4.5 Origin of correlations based on a simple hidden variable theory.

tion. In fact, we can see immediately that the properties we have ascribed to the hidden variables will not allow joint $+ -$ or $- +$ results, and so this theory is entirely consistent with quantum theory for this particular arrangement of the polarization analysers.

Now let us rotate PA_2 through some angle $(b - a)$ with respect to PA_1. We denote the new polarization axes of PA_2 as v' and h'. Again we assume for the sake of simplicity that photon A gives a $+$ result. This has the same implications for the hidden variable of photon B as before, i.e. λ points in the opposite direction and lies within $\pm 45°$ of the v axis. However, for photon B to give a $+$ result, λ must lie within $\pm 45°$ of the new v' axis (see Fig. 4.5). Clearly the joint probability $P_{++}(a, b)$ will depend on the probability that λ for photon B lies within $\pm 45°$ of *both* the v and v' axes — the doubly-shaded area shown in Fig. 4.5. This probability is given by the ratio of the range of angles that determine the area of overlap $(90° - |b - a|)$ to the range of all possible angles $(180°)$. Thus,

$$P_{++}(a, b) = (90° - |b - a|)/180°. \qquad (4.20)$$

This expression for $P_{++}(a, b)$ is valid for $0° \leqslant |b - a| \leqslant 90°$. We can use a similar line of reasoning to show that

$$P_{+-}(a, b) = |b - a|/180°$$
$$P_{-+}(a, b) = |b - a|/180° \qquad (4.21)$$
$$P_{--}(a, b) = (90° - |b - a|)/180°.$$

From eqn (4.18), it follows that the prediction for $E(a,b)$ from this simple local hidden variable theory is

$$E(a, b) = (180° - 4|b - a|)/180°$$
$$= 1 - |b - a|/45° \qquad (4.22)$$

valid for $0° \leqslant |b - a| \leqslant 90°$.

Note that when $|b - a| = 0°$, 45° and 90°, $E(a, b) = +1$, 0 and -1 respectively. This local hidden variable theory is therefore consistent with the quantum theory predictions at these three angles. However, from the comparison of the two correlation functions shown in Fig. 4.6, we can see that the two theories predict different results at all other angles. The greatest difference between them occurs at $(b - a) = 22.5°$, where quantum theory predicts $E(a,b) = \cos 45° = 1/\sqrt{2}$ and the local hidden variable theory predicts $E(a,b) = \frac{1}{2}$.

This appears to be merely a confirmation of Bohm's contention, quoted above, that 'no theory of mechanically determined hidden variables can lead to *all* the results of the quantum theory.' But you might not yet be satisfied that the case is proven. After all, the local hidden variable theory we have described here is a very simple one. Might it not be possible to devise a more complicated version that could reproduce all the results of quantum theory? More complicated local hidden variable theories are indeed possible, but, in fact, *none* can reproduce all the

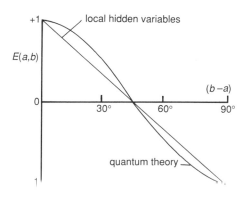

Fig. 4.6 Comparison of the dependences of the quantum theory and hidden variables correlations on the angle between the vertical axes of the analysers.

predictions of quantum theory. The truth of this statement is demonstrated in a celebrated theorem devised by John S. Bell.

4.3 BELL'S THEOREM

Bohm's early work on the EPR experiment and non-local hidden variables reawakened the interest of a small section of the physics community in these problems. Many dismissed Bohm's work as 'old stuff, dealt with long ago', but for some his approach served to heighten their own unease about the interpretation of quantum theory, even if they did not necessarily share his conclusions. One physicist who became very suspicious was John S. Bell. In a paper submitted to the journal *Reviews of Modern Physics* in 1964 (but not actually published until 1966), Bell examined, and rejected, von Neumann's 'impossibility proof' and similar arguments that had been used to deny the possibility of hidden variables.

However, in a subsequent paper, Bell demonstrated that under certain conditions quantum theory and local hidden variable theories predict different results for the same experiments on pairs of correlated particles. This difference, which is intrinsic to *all* local hidden variable theories and is independent of the exact nature of the theory, is summarized in Bell's theorem. Questions about local hidden variables immediately changed character. From being rather academic questions about philosophy they became practical questions of profound importance for quantum theory. The choice between quantum theory and local hidden variable theories was no longer a matter of taste, it was a matter of *correctness*.

Bertlmann's socks

We will derive Bell's theorem through the agency of Dr Bertlmann, a real character used by Bell for a discussion on the nature of reality which was published in the *Journal de Physique* in 1981. I can find no better introduction than to use Bell's own words:[†]

The philosopher in the street, who has not suffered a course in quantum mechanics, is quite unimpressed by Einstein–Podolsky–Rosen correlations. He can point to many examples of similar correlations in everyday life. The case of Bertlmann's socks is often cited. Dr Bertlmann likes to wear two socks of different colours. Which colour he will have on a given foot on a given day is quite unpredictable. But when you see [Fig. 4.7] that the first sock is pink you can be

[†] Bell, J.S. (1981) *Journal de Physique*. Colloque C2, suppl. au numero 3, tome 42.

Les chaussettes
de M. Bertlmann
et la nature
de la réalité

Foundation Hugot
juin 17 1980

pink

not
pink →

Fig. 4.7 Bertlmann and the nature of reality. Reprinted with permission from *Journal de Physique (Paris), Colloque C2*, (suppl. au numero 3), **42** (1981) C2 41–61.

already sure that the second sock will not be pink. Observation of the first, and experience of Bertlmann, gives immediate information about the second. There is no accounting for tastes, but apart from that there is no mystery here. And is not this EPR business just the same?

Dr Bertlmann happens to be a physicist who is very interested in the physical characteristics of his socks. He has secured a research contract from a leading sock manufacturer to study how his socks stand up to the rigours of prolonged washing at different temperatures. Bertlmann decides to subject his left socks (socks A) to three different tests:

test a, washing for 1 hour at 0 °C;
test b, washing for 1 hour at 22.5 °C;
test c, washing for 1 hour at 45 °C.

He is particularly concerned about the numbers of socks A that survive intact (+ result) or are destroyed (− result) by prolonged washing at these different temperatures. He denotes the number of socks that pass test a and fail test b as $n[a_+ b_-]$. Being a theoretical physicist, he knows that he can discover some simple relationships between such numbers without actually having to perform the tests using real socks and real washing machines. This makes his study inexpensive and therefore attractive to his research sponsors.

He reasons that $n[a_+ b_-]$ can be written as the sum of the numbers of socks which belong to two subsets, one in which the individual socks pass test a, fail b and pass c and one in which the socks pass test a, fail b and fail c:

$$n[a_+b_-] = n[a_+b_-c_+] + n[a_+b_-c_-].\tag{4.23}$$

Similarly,

$$n[b_+c_-] = n[a_+b_+c_-] + n[a_-b_+c_-]\tag{4.24}$$

and

$$n[a_+c_-] = n[a_+b_+c_-] + n[a_+b_-c_-].\tag{4.25}$$

From eqn (4.23) it follows that

$$n[a_+b_-] \geqslant n[a_+b_-c_-]\tag{4.26}$$

and from eqn (4.24) it follows that

$$n[b_+c_-] \geqslant n[a_+b_+c_-].\tag{4.27}$$

Adding eqns (4.26) and (4.27) gives

$$n[a_+b_-] + n[b_+c_-] \geqslant n[a_+b_-c_-] + n[a_+b_+c_-]\tag{4.28}$$

or

$$n[a_+b_-] + n[b_+c_-] \geqslant n[a_+c_-].\tag{4.29}$$

It is at this stage that Bertlmann notices the flaw in his reasoning which readers will, of course, have spotted right at the beginning. Subjecting one of the socks A to test a will necessarily change irreversibly its physical characteristics such that, even if it survives the test, it may not give the result for test b that might be expected of a brand new sock. And, of course, if the sock fails test b, it will simply not be available for test c. The numbers $n[a_+b_-]$ etc therefore have no *practical* relevance.

But then Bertlmann remembers that his socks always come in *pairs*. He assumes that, apart from differences in colour, the physical characteristics of each sock in a pair are identical. Thus, a test performed on the right sock (sock B) can be used to predict what the result of the same test would be if it was performed on the left sock (sock A), even though the test on A is not actually carried out. He must further assume that whatever test he chooses to perform on B in no way affects the outcome of any other test he might perform on A, but this seems so obviously valid that he does not give it a second thought.

Bertlmann now devises three different sets of experiments to be carried out on three samples containing the same total number of pairs of his socks. In experiment 1, for each pair, sock A is subjected to test a and sock B is subjected to test b. If sock B fails test b, this implies that sock

A would also have failed test b had it been performed on A. Thus, the number of pairs of socks for which A passes test a and B fails test b, $N_{+-}(a, b)$, must be equal to the (hypothetical) number of socks A which pass test a and fail test b, i.e

$$N_{+-}(a,b) = n[a_+ b_-]. \qquad (4.30)$$

In experiment 2, for each pair, sock A is subjected to test b and sock B is subjected to test c. The same kind of reasoning allows Bertlmann to deduce that

$$N_{+-}(b,c) = n[b_+ c_-]. \qquad (4.31)$$

Finally, in experiment 3, for each pair, sock A is subjected to test a and sock B is subjected to test c. Bertlmann deduces that

$$N_{+-}(a,c) = n[a_+ c_-]. \qquad (4.32)$$

The arrangements for each experiment are conveniently summarized below.

Experiment	Test	
	Sock A	Sock B
1	a	b
2	b	c
3	a	c

From eqns (4.30)–(4.32) and (4.29) Bertlmann has, therefore

$$N_{+-}(a, b) + N_{+-}(b, c) \geqslant N_{+-}(a, c). \qquad (4.33)$$

Bertlmann now generalizes this result for any batch of pairs of socks. By dividing each number in eqn (4.33) by the total number of pairs of socks (which was the same for each experiment) he arrives at the frequencies with which each joint result was obtained. He identifies these frequencies with probabilities for obtaining the results for experiments to be performed on any batch of pairs of socks that, statistically, have the same properties. Thus,

$$P_{+-}(a, b) + P_{+-}(b, c) \geqslant P_{+-}(a, c). \qquad (4.34)$$

This is Bell's inequality.

Bell's inequality

While this digression has been entertaining, readers might be wondering about its relevance to quantum physics. Actually, it is very relevant.

Follow the above arguments through once more, replacing socks with photons, pairs of socks with pairs of correlated photons, washing machines with polarization analysers and temperatures with polarizer orientations and you will still arrive at Bell's inequality, eqn (4.34).

Our three tests now refer to polarization analysers set with their vertical axes oriented at $a = 0°$, $b = 22.5°$ and $c = 45°$. These different arrangements can be summarised as follows:

Experiment	Photon A PA$_1$ orientation	Photon B PA$_2$ orientation	Difference
1	a (0°)	b (22.5°)	$b - a = 22.5°$
2	b (22.5°)	c (45°)	$c - b = 22.5°$
3	a (0°)	c (45°)	$c - a = 45°$

The expressions in eqn (4.16) give the probabilities as predicted by quantum theory for any angle $(b - a)$. Putting in the appropriate angles allows us to rewrite eqn (4.34) as follows

$$\frac{1}{2} \sin^2(22.5°) + \frac{1}{2} \sin^2(22.5°) \geqslant \frac{1}{2} \sin^2(45°) \qquad (4.35)$$

or

$$0.1464 \geqslant 0.2500 \qquad (4.36)$$

which is obviously incorrect. Thus, for these particular arrangements of the polarization analysers, quantum theory predicts results that violate Bell's inequality.

The most important assumption we made in the reasoning which led to this inequality was that of Einstein separability or local reality of the photons. It is therefore an inequality that is quite independent of the nature of any local hidden variable theory that we could possibly devise. The conclusion is inescapable, quantum theory is incompatible with *any* local hidden variable theory and hence local reality. (Readers might wish to confirm for themselves that the simple local hidden variable theory described above, for which the predicted probabilities are given in eqn (4.21), does indeed conform to Bell's inequality for the same set of angles.)

We should not, perhaps, be too surprised by this result. The predictions of quantum theory are based on the properties of a two-particle state vector which, before collapsing into one of the measurement eigenstates, is 'delocalized' over the whole experimental arrangement. The two particles are, in effect, always in 'contact' prior to measurement and can therefore exhibit a degree of correlation that is impossible for two Einstein separable particles. However, Bell's inequality provides us with a straightforward test. If experiments like the ones described here are actually performed, the results will allow us to choose between quantum theory and a whole range of theories based on local hidden variables.

Generalization

Before we get too carried away with these inequalities, we should remember what it is we are supposed to be measuring here. We are proposing an experiment in which some atomic source (yet to be specified) emits a pair of photons correlated so that they have no net angular momentum. The photons move apart and each enter a polarization analyser oriented at some angle to the arbitrary laboratory vertical axis. The photons are detected to emerge from the vertical or horizontal polarization channels of these analysers and the results of coincident measurements are compared with the predictions of quantum theory and local hidden variable theories.

Unfortunately, nothing in this life is ever easy. Bertlmann's derivation of the inequality (4.34) is based on an important assumption. Remember that he had supposed that, with the exception of colour, each member of any given pair of his socks possesses identical physical characteristics so that the result of any test performed on sock B would automatically imply the same result for A. This, in turn, implies that if we perform the same test on both socks simultaneously, we expect to observe identical results, or perfect correlation. In the language of the equivalent experiments with photons, if we orient PA_1 and PA_2 so that their vertical axes are parallel, we expect to obtain perfect correlation — $E(a, a) = +1$, $P_{+-}(a, a) = P_{-+}(a, a) = 0$. Alas, in the 'real' world, there are a number of limitations in the experimental technology of polarization measurements that prevent us from observing perfect correlation. And any effect that reduces the physicist's ability to measure these correlations below the maximum permitted by Bell's inequality will render the experiments inconclusive.

Firstly, real polarization analysers are not 'perfect'. They do not transmit all the photons that are incident on them (through one or other of the two channels) and they often 'leak', i.e. horizontally polarized photons can occasionally pass through the vertical channel, and vice versa. Worse still, the transmission characteristics of the analysers may depend on their orientation. Secondly, detectors such as photomultipliers are quite inefficient, producing measurable signals for only a small number of the photons actually generated. Finally, the analysers and detectors themselves must be of limited size, and so they cannot 'gather' all of the photons emitted, even if they are emitted in roughly the right direction. Experimental factors such as these limit the numbers of pairs that can be detected successfully, and will also lead to some pairs being detected 'incorrectly'; for example, a pair which should have given a + + result actually being recorded as a + − result. These limitations always serve to reduce the extent of correlation between the photons that can be observed experimentally.

There is a way out of this impasse. It involves a generalization of Bell's inequality to include a fourth experimental arrangement, and was first derived by John F. Clauser, Michael A. Horne, Abner Shimony and Richard A. Holt in a paper published in *Physical Review Letters* in 1969. A derivation is provided in Appendix B.

Denoting the four different orientations of the polarization analysers as *a, b, c* and *d* this generalized form of Bell's inequality can be written:

$$|E(a, b) - E(a, d)| + |E(c, b) + E(c, d)| \leqslant 2. \qquad (4.37)$$

The advantage of this generalization is that nowhere in its derivation is it necessary to rely on perfect correlation between the measured results for any combination of polarizer orientations (see Appendix B). Inequality (4.37) applies equally well to non-ideal cases. For future convenience, we denote the term involving the different expectation values on the left-hand side of eqn (4.37) by the symbol S. We will use different subscripts to differentiate between theoretical predictions for and experimental measurements of S.

There is a further important point of which we should take note. The implication of the hidden variable approach we have so far adopted is that the λ values are set at the moment the photons are emitted, and the outcomes of the measurements therefore predetermined. However, there is nothing in the derivation of eqn (4.37) which says this must be so. The only assumption needed is one of *locality* — measurements made on photon A do not affect the possible outcomes of any subsequent measurements made on B and vice versa. The generalized form of Bell's inequality actually provides a test for all classes of locally realistic theories, not just those theories which happen also to be deterministic. It is no longer essential to suppose that the λ values of photons A and B remain determined as they propagate towards their respective analysers.

The photons must still be correlated (no net angular momentum) but their λ values could vary between emission and detection. All that is required for eqn (4.37) to be valid is that there should be no communication between the photons at the moment a measurement is made on one of them. As we can arrange for the analysers to be a long distance apart (or space-like separated, to use the physicists' term) this requirement essentially means no communication faster than the speed of light. Recall once again that Einstein suspected that the Copenhagen interpretation of quantum theory might necessarily lead to a violation of the postulates of special relativity.

Now let us take a look at some specific orientations in actual experiments. Consider the four experimental arrangements summarized below:

Experiment	Photon A PA$_1$ orientation	Photon B PA$_2$ orientation	Difference
1	a (0°)	b (22.5°)	$b - a = 22.5°$
2	a (0°)	d (67.5°)	$d - a = 67.5°$
3	c (45°)	b (22.5°)	$b - c = -22.5°$
4	c (45°)	d (67.5°)	$d - c = 22.5°$

From eqn (4.19), we note that the expectation value for the joint measurement with the vertical axes of the polarization analysers at an angle $(b - a)$ is $\cos 2(b - a)$. Thus,

$$S_{QT} = |E(a, b) - E(a, d)| + |E(c, b) + E(c, d)|$$
$$= |\cos(45°) - \cos(135°)| + |\cos(-45°) + \cos(45°)|$$
$$= |\frac{1}{\sqrt{2}} + \frac{1}{\sqrt{2}}| + |\frac{1}{\sqrt{2}} + \frac{1}{\sqrt{2}}|$$
$$= 2\sqrt{2} = 2.828. \tag{4.38}$$

where S_{QT} denotes the quantum theory prediction for S. Equation (4.38) is in clear violation of inequality (4.37). Readers can once again satisfy themselves that the simple local hidden variable theory described in Section 4.2, which predicts $E(a, b) = 1 - |b - a|/45°$ for $0° \leqslant |b - a| \leqslant 90°$, further predicts $S_{HV} = 2$, where the subscript HV stands for 'hidden variables'.

This exercise merely confirms once more that quantum theory is not consistent with local reality. Correlations between the photons can be greater than is possible for two Einstein separable particles since the reality of their physical properties is not established until a measurement is made. The two particles are in 'communication' over large distances since their behaviour is governed by a common state vector. Quantum theory demands a 'spooky action at a distance' that violates special relativity.

4.4 THE ASPECT EXPERIMENTS

It is probably reasonable to suppose that the derivation, in the late 1960s and early 1970s, of an equation which is demonstrably violated by a quantum theory then over 40 years old should have settled the matter one way or the other, once and for all. Correlated quantum particles are everywhere in physics and chemistry, the simplest and most obvious example being the helium atom, an understanding of the spectroscopy of which had led to the introduction of the Pauli principle in the first place. But it became apparent that the special circumstances under which Bell's inequality could be subjected to experimental test had never been realized in the laboratory. Suddenly, the race was on to perfect an apparatus that could be used to perform the necessary measurements on pairs of correlated quantum particles.

As early as 1946, the physicist John Wheeler, then at Princeton University, had proposed studies on correlated photons produced by electron–positron annihilation. But the polarization correlations of two photons emitted in rapid succession (in a 'cascade') from an excited state of the calcium atom proved to be the most accessible to experiment and ultimately closest to the ideal. Carl A. Kocher and Eugene D. Commins at the University of California at Berkeley used this source in 1966 in a study of correlations between the linear polarization states of the photons, although they did not explicitly set out to test Bell's inequality.

The first such direct tests were performed in 1972, by Stuart J. Freedman and John F. Clauser at the Lawrence Livermore Laboratory in California, who also used the calcium atom source. These experiments produced the violations of Bell's inequality predicted by quantum theory but, because of some further 'auxiliary' assumptions that were necessary in order to extrapolate the data, only a weaker form of the inequality was tested. These auxiliary assumptions left unsatisfactory loopholes for the ardent supporters of local hidden variables to exploit. It could still be argued then that the evidence against such hidden variables was only circumstantial.

To date, the best, most comprehensive experiments designed specifically to test the general form of Bell's inequality were those performed by Alain Aspect and his colleagues Philippe Grangier, Gérard Roger and Jean Dalibard, at the Institut d'Optique Théoretique et Appliquée, Université Paris-Sud in Orsay, in 1981 and 1982. These scientists also made use of cascade emission from excited calcium atoms as a source of correlated photons. We will now examine the physics of this emission process in detail.

Cascade emission

In the lowest energy (ground) state of the calcium atom, the outermost 4s orbital is filled with two spin-paired electrons. The vector sum of the spin angular momenta of these electrons is therefore zero, and the state is characterized by a total spin quantum number $S = 0$. The spin multiplicity $(2S + 1)$ is unity and so the state is called a singlet state.

The total angular momentum of the atom is a combination of the intrinsic angular momentum that the electrons possess by virtue of their spins and the angular momentum they possess by virtue of their orbital motion. We can combine these two kinds of angular momentum in different ways. In the first, we determine separately the total spin angular momentum (characterized by the quantum number S) and the total orbital angular momentum (quantum number L) and combine these to give the overall momentum (quantum number J). In the second we combine the spin and orbital angular momenta of each individual electron (quantum number j) and combine these to give the overall total. The former method is appropriate for atoms with light nuclei and we will use it here.

In fact, for the ground state of the calcium atom, the outermost electrons are both present in a spherically symmetric s orbital and therefore possess no orbital angular momentum: $L = 0$, $S = 0$ and so $J = 0$. The state is labelled $4s^2\ ^1S_0$, where the superscript 1 indicates that it is a singlet state, the S indicates that $L = 0$ (S corresponds to $L = 0$, P corresponds to $L = 1$, D corresponds to $L = 2$, etc) and the subscript 0 indicates that $J = 0$.

If we use light to excite the ground state of a calcium atom, the photon that is absorbed imparts a quantum of angular momentum to the atom. This extra angular momentum cannot appear as electron spin, since that is fixed at $\frac{1}{2}\hbar$. Thus, the angular momentum must appear in the excited electron's orbital motion, and so the value of L must increase by one unit. Promoting one electron from the 4s orbital to the 4p orbital satisfies this selection rule. If there is no change in the spin orientations of the electrons, the excited state is still a singlet state, $S = 0$ and, since $L = 1$, there is only one possible value for J, $J = 1$. This excited state is labelled $4s4p\ ^1P_1$.

Now imagine that we could somehow excite a second electron (the one left behind in the 4s orbital) also into the 4p orbital, but still maintaining the alignment of the electron spins. The configuration would then be $4p^1$, which can give rise to three different electronic states corresponding to the three different ways of combining the two orbital angular momentum vectors. In one of these states the orbital angular momentum vectors of the individual electrons cancel, $L = 0$ and, since $S = 0$ we have $J = 0$. This particular doubly excited state is labelled $4p^2\ {}^1S_0$.

If this doubly excited state is produced in the laboratory, it undergoes a rapid cascade emission through the $4s4p\ {}^1P_1$ state to return to the ground state (see Fig. 4.8). Two photons are emitted. Because the quantum number J changes from $0 \rightarrow 1 \rightarrow 0$ in the cascade, the net angular momentum of the photon pair must be zero. The photons are therefore emitted in opposite states of circular polarization. In fact, the photons have wavelengths in the visible region. Photon A, from the $4p^2\ {}^1S_0 \rightarrow 4s4p\ {}^1P_1$ transition, has a wavelength of 551.3 nm (green) and photon B, from the $4s4p\ {}^1P_1 \rightarrow 4s^2\ {}^1S_0$ transition, has a wavelength of 422.7 nm (blue).

Experimental details

In the experiments conducted by Aspect and his colleagues, the $4p^2\ {}^1S_0$ state was not produced by the further excitation of the $4s4p\ {}^1P_1$ state, since that would have required light of the same wavelength as photon B, making isolation and detection of the subsequently emitted light very difficult. Instead, the scientists used two high-power lasers, with wavelengths of 406 and 581 nm, to excite the calcium atoms. The very high intensities of lasers make possible otherwise very low probability multiphoton excitation. In this case two photons, one of each colour, were absorbed simultaneously by a calcium atom to produce the doubly excited state (see Fig. 4.8).

Aspect, Grangier and Roger actually used a calcium atomic beam. This was produced by passing gaseous calcium from a high temperature oven through a tiny hole into a vacuum chamber. Subsequent collimation of the atoms entering the sample chamber provided a well defined beam of atoms with a density of about 3×10^{10} atoms cm^{-3} in the

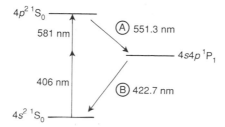

Fig. 4.8 Electronic states of the calcium atom involved in two-photon cascade emission process.

region where the atomic beam intersected the laser beams. This low density (atmospheric pressure corresponds to about 2×10^{19} molecules cm^{-3}) ensured that the calcium atoms did not collide with each other or with the walls of the chamber before absorbing and subsequently emitting light. It also removed the possibility that the emitted 422.7 nm light would be reabsorbed by ground state calcium atoms.

Figure 4.9 is a schematic diagram of the apparatus used by Aspect and his colleagues. They monitored light emitted in opposite directions from the atomic beam source, using filters to isolate the green photons (A) on the left and the blue photons (B) on the right. They used two polarization analysers, four photomultipliers and electronic devices designed to detect and record coincident signals from the photomultipliers. The polarization analysers were actually polarizing cubes, each made by gluing together two prisms with dielectric coatings on those faces in contact. These cubes transmitted light polarized parallel (vertical) to the plane of incidence, and reflected light polarized perpendicular (horizontal) to this plane. Thus, detection of a transmitted photon corresponds in our earlier discussion to a + result, while detection of a reflected photon corresponds to a − result.

The polarizing cubes were neither quite perfectly transmitting for pure vertically polarized light nor perfectly reflecting for pure horizontally polarized light. The physicists measured the transmittance of PA_1 for

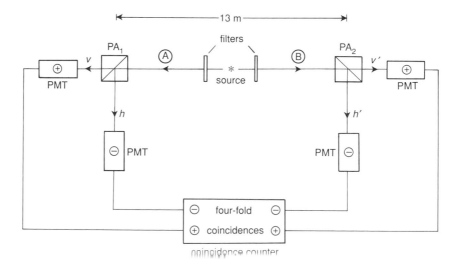

Fig. 4.9 Schematic diagram of the experimental apparatus used by Aspect, Grangier and Roger (PMT is a photomultiplier).

vertically polarized light, T_1^v, and the reflectance of PA_1 for horizontally polarized light, R_1^h for light with a wavelength of 551.3 nm. They obtained $T_1^v = R_1^h = 0.950$. They also measured $T_1^h = R_1^v = 0.007$. These latter figures represent a small amount of 'leakage' through the analyser. Similarly, for light with a wavelength of 442.7 nm, they measured $T_2^v = R_2^h = 0.930$ and $T_2^h = R_2^v = 0.007$.

Each polarization analyser was mounted on a platform which allowed it to be rotated about its optical axis. Experiments could therefore be performed for different relative orientations of the two analysers. The analysers were placed about 13 m apart. The electronics were set to look for coincidences in the arrival and detection of the photons A and B within a 20 nanosecond (ns) time window. This is large compared with the time taken for the intermediate $4s4p\ ^1P_1$ state to decay (about 5 ns), and so all true coincidences were counted.

Note that to be counted as a coincidence, the photons had to be detected within 20 ns of each other. Any kind of signal passed between the photons, 'informing' photon B of the fate of photon A, for example, must therefore have travelled the 13 m between the analysers and detectors within 20 ns. In fact, it would take about 40 ns for a signal moving at the speed of light to travel this distance. The two analysers were therefore space-like separated.

The results

Aspect, Grangier and Roger actually measured coincidence *rates* (coincidences per unit time). For the specific arrangement in which PA_1 has orientation a and PA_2 has orientation b, we write these coincidence rates as $R_{++}(a, b)$, $R_{+-}(a, b)$, $R_{-+}(a, b)$ and $R_{--}(a, b)$. After correction for accidental coincidences, the physicists obtained results which varied in the range 0–40 coincidences s^{-1} depending on the angle between the vertical axes of the polarisers $(b - a)$. They then used these results to derive an experimental expectation value, $E(a, b)_{expt}$, for comparison with theory (cf. eqn (4.18)):

$$E(a, b)_{expt} = \frac{R_{++}(a, b) - R_{+-}(a, b) - R_{-+}(a, b) + R_{--}(a, b)}{R_{++}(a, b) + R_{+-}(a, b) + R_{-+}(a, b) + R_{--}(a, b)}.$$

$$(4.39)$$

Dividing by the sum of the coincidence rates normalizes the expectation value (it is equivalent to dividing by the total number of photon pairs detected).

The physicists measured the expectation value for seven different sets of analyser orientations, and the results they obtained are shown in Fig. 4.10. From eqn (4.19), we know that the quantum theory prediction

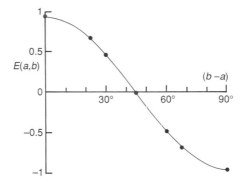

Fig. 4.10   Results of measurements of the expectation value $E(a, b)$ for seven different relative orientations of the polarization analysers. The continuous line represents the quantum theory predictions modified to take account of instrumental factors (see text). Reprinted with permission from Aspect *et al.* (1982). *Physical Review Letters*, **49**, 91.

for $E(a, b)$ is $\cos 2(b - a)$. However, the extent of the correlation observed experimentally was dampened by limitations in the apparatus, as described above. The physicists therefore derived a slightly modified form of the quantum theory prediction that takes these limiting factors into account. They obtained:

$$E(a, b) = F \frac{(T_1^v - T_1^h)(T_2^v - T_2^h)}{(T_1^v + T_1^h)(T_2^v + T_2^h)} \cos 2(b - a). \qquad (4.40)$$

The factor F allows for the finite solid angles for detection of the photons (not all photons could be physically 'gathered' into the detection system). They found $F = 0.984$ for their experimental arrangement. The term involving the analyser transmittances accounts for the small amount of leakage and for the fact that not all photons incident on the analysers were ultimately detected. The *predictions* of quantum theory, corrected for these instrumental deficiencies, are shown in Fig. 4.10 as the continuous line. As expected, the predictions demonstrate that perfect correlation, $(b - a) = 0°$, and perfect anticorrelation, $(b - a) = 90°$, were not quite realized in these experiments.

Aspect and his colleagues then performed four sets of measurements with analyser orientations as described on p. 138. Defining the quantity S_{expt} as $[|E(a, b) - E(a, d)| + |E(c, b) + E(c, d)|]_{expt}$ (cf. eqn (4.38)), from their measurements they obtained

$$S_{expt} = 2.697 \pm 0.015 \qquad (4.41)$$

a violation of Bell's inequality, eqn (4.37)), by 83% of the maximum possible predicted by quantum theory (i.e. $\sqrt{2}$, see eqn (4.38)). By taking the instrumental limitations into account, the physicists obtained a modified quantum theory prediction for this quantity of $S_{QT} = 2.70 \pm 0.05$, in excellent agreement with experiment.

These results provide almost overwhelming evidence in favour of quantum theory against all classes of locally realistic theories. One loophole remained, however. The polarization analysers were set in position *before* the experiments were initiated (i.e. before the calcium atoms were excited and, most importantly, before the correlated photons were emitted). Could it not be that the photons were somehow influenced in advance by the way the apparatus was set up? If so, is it possible that the photons could have been emitted with just the right physical characteristics (governed, of course, by local hidden variables) to reproduce the quantum theory correlations? Although this is beginning to look like some kind of grand conspiracy on the part of the photons, it is not a possibility that can be excluded by the experiments just described.

Closing the last loophole

To close this last remaining loophole, Aspect, Dalibard and Roger modified the experimental set-up to include two acousto-optical switching devices (see Fig. 4.11). Each device was designed to switch the incoming photons rapidly between two different optical paths, and each was activated by passing standing ultrasonic waves through a small volume of water held in a transparent container. The ultrasonic waves, which change the refractive index of the water and hence change the path of light passing through it, were driven at frequencies designed to switch between the two paths every 10 ns. At the end of each path was placed a polarization analyser (which could be oriented independently of the

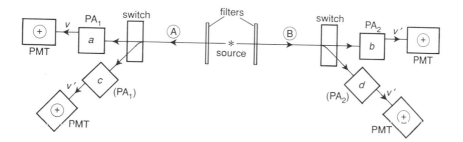

Fig. 4.11 Schematic diagram of the experimental apparatus used by Aspect, Dalibard and Roger.

rest of the apparatus) and a photomultiplier. The vertical axes of each of the four analysers were oriented in different directions.

By this arrangement, the physicists prevented the photons from 'knowing' in advance along which optical path they would be travelling, and hence through which analyser they would eventually pass. The end result was equivalent to changing the relative orientations of the two analysers while the photons were in flight. Any communication between the photons regarding the way the apparatus was set up was therefore restricted to the moment of measurement, in principle requiring faster than light signalling between the photons to establish the correlation.

The switching arrangement shown in Fig. 4.11 was difficult to operate and run successfully. Aspect and his colleagues could not add to this difficulty by trying to detect photons both transmitted and reflected by the polarization analysers (such an experiment would have required eight photomultipliers and the necessary coincidence detection!). The physicists could therefore only detect those photons transmitted by the analysers: they could observe only + results. Fortunately, a version of Bell's inequality had been derived by John F. Clauser, Michael A. Horne, Abner Shimony and Richard A. Holt in 1969 for just this kind of experiment. We will derive this inequality here.

Let us return for a moment to the expression for the expectation value $E(a, b)$ given in eqn (4.18):

$$E(a, b) = P_{++}(a, b) - P_{+-}(a, b) - P_{-+}(a, b) + P_{--}(a, b).$$
(4.18)

Our difficulty arises because if only transmitted photons are detected, then only quantities related to $P_{++}(a, b)$ can be measured. However, consider the equivalent experiment performed with PA$_2$ removed completely. We use the symbol ∞ instead of b to define a probability for joint detection, $P_{++}(a, \infty)$, in these circumstances. Provided the removal of PA$_2$ in no way affects the behaviour of either photon A or B, then $P_{++}(a, \infty)$ should include the probabilities of all possible joint results in which photon A is detected, i.e. it includes the possible joint results in which photon B is detected $(+)$ and not detected $(-)$:

$$P_{++}(a, \infty) = P_{++}(a, b) + P_{+-}(a, b).$$
(4.42)

Similarly,

$$P_{++}(\infty, b) = P_{++}(a, b) + P_{-+}(a, b)$$
(4.43)

and

$$P_{++}(\infty, \infty) = P_{++}(a, b) + P_{+-}(a, b) + P_{-+}(a, b) + P_{--}(a, b).$$
(4.44)

We can now combine these expressions to give

$$E(a, b) = 4P_{++}(a, b) - 2P_{++}(a, \infty) - 2P_{++}(\infty, b) + P_{++}(\infty, \infty).$$
(4.45)

Equation (4.45) allows us to calculate the expectation value using only the probabilities of joint + + results, for which related quantities can be obtained from experiment.

From eqn (4.45) it follows that

$$E(a, b) - E(a, d) = 4P_{++}(a, b) - 4P_{++}(a, d)$$
$$- 2P_{++}(\infty, b) + 2P_{++}(\infty, d)$$
(4.46)

and

$$E(c, b) + E(c, d) = 4P_{++}(c, b) + 4P_{++}(c, d) - 4P_{++}(c, \infty)$$
$$- 2P_{++}(\infty, b) - 2P_{++}(\infty, d) + 2P_{++}(\infty, \infty).$$
(4.47)

Equations (4.46) and (4.47) can now be combined to give an expression for S in terms of the probabilities for joint + + results.

In the experiments, coincidence rates were actually measured. When normalized, these rates are related to the joint detection probabilities via relations such as

$$P_{++}(a, b) = \frac{R_{++}(a, b)}{R_{++}(\infty, \infty)}.$$
(4.48)

All the quantities needed to obtain S from the experiments with switched optical paths were measured by Aspect and his colleagues. For $a = 0°$, $b = 22.5°$, $c = 45°$ and $d = 67.5°$, they obtained $S_{expt} = 2.404 \pm 0.080$, once again in clear violation of inequality (4.37). Taking account of inefficiencies in the polarization analysers and the finite solid angles for detection allowed them to obtain a modified quantum theory prediction $S_{QT} = 2.448$, in excellent agreement with experiment.

So, where does all this leave local reality? For the purist, the last loophole is still not completely closed by these experiments. The standing ultrasonic waves used to drive the acousto-optical switches did not provide completely random switching, although the two switches were driven at different frequencies. However, we would need to invoke a very grand conspiracy indeed to salvage local hidden variables from these experimental results. This immediately brings to mind another of Einstein's famous quotes (made in a rather different context): 'The Lord is subtle, but he is not malicious.'

The majority of physicists, including those like David Bohm and John Bell who have rejected the Copenhagen view, have accepted that the

Aspect experiments create great difficulties for theories which feature a local reality. Either we give up reality or we accept that there can be some kind of 'spooky action at a distance', involving communication between distant parts of the world at speeds faster than that of light. This appears to conflict with the postulates of special relativity.

Although the independent reality advocated by the realist does not have to be a local reality, it is clear that the experiments described here leave the realist with a lot of explaining to do. An observer changing the orientation of a polarizer *does* affect the behaviour of a distant photon, no matter how distant it is. Whatever the nature of reality, it cannot be as simple as we might have thought at first.

Do the Aspect experiments necessarily represent the end of this story as far as experimental physics is concerned? In 1985, Bell thought not:[†]

It is a very important experiment, and perhaps it marks the point where one should stop and think for a time, but I certainly hope it is not the end. I think that the probing of what quantum mechanics means must continue, and in fact it will continue, whether we agree or not that it is worth while, because many people are sufficiently fascinated and perturbed by this that it will go on.

Superluminal communications

Whether or not we accept that correlated photons are objectively real entities which exist independently of our instruments, the results of the Aspect experiments suggest an interesting possibility. Can we exploit the communication that seems to take place between distant photons to send faster-than-light messages? To answer this question we need to devise a simple procedure by which information might be communicated between two distant observers, and then see if such a procedure works in principle.

The feature of the physics of the correlated photons that we must try to exploit is the instantaneous realization of a specific polarization state for photon B at the moment that photon A is detected to be in a specific polarization state. Consider the AMAZING Superluminal Communications System™, manufactured and marketed by the AMAZING Company of Reading, U.K., shown schematically in Fig. 4.12. It has three parts, a transmitter, a receiver, and a 'line' provided by a central source of correlated photons emitted continuously in opposite directions, at regular intervals of say 10 ns. The photons that make up a pair are timed to arrive coincidentally at the transmitter and receiver.

[†] Bell, J.S. in Davies, P.C.W. and Brown, J.R. (1986). *The ghost in the atom*. Cambridge University Press.

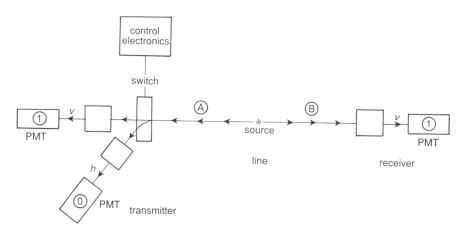

Fig. 4.12   The AMAZING Superluminal Communications System™.

The transmitter is built around an acousto-optical switch which switches the incoming A photons between two different optical paths. One path leads to a polarizing filter oriented with its axis of maximum transmission in the vertical direction and the other leads to a polarizing filter oriented horizontally. The transmitter electronics recognizes detection of a photon through the vertical polarizer as a '1', and detection of a photon through the horizontal polarizer as a '0'. The receiver, located on the moon, has only one polarization filter, oriented vertically.

Now suppose that we wish to inform a friend on the moon that the temperature in Reading is currently 19 °C. This number can be encoded as a binary number (in fact, the binary form of 19 is 10011). We plug this binary number into the electronic system that controls the acousto-optical switch. When the system wants to send a 1, the next incoming A photon is switched through to the vertical polarizer. Its detection forces photon B into a vertical polarization state (because of the correlation between the photons $- E(a, a) = +1$), which passes through the vertical polarizer in the receiver and is detected. This signal is recognized by the receiver electronics as a 1 and the digit has therefore been transmitted instantaneously.

When the transmitter wants to send a 0, the next incoming A photon is switched through to the horizontal polarizer. Its detection forces the next photon B into a horizontal polarization state which is blocked by the polarizer in the receiver. Since the receiver expects the next photon within 10 ns of the previous one, it recognizes non-detection as a 0.

This process continues until all the binary digits have been sent. Our distant friend decodes the binary number and learns that the temperature in Reading is 19 °C. The information takes about 50 ns to transmit, compared to the 1.3 s or so that it takes a conventional signal to travel the 240 000 miles from the earth to the moon. This represents a time saving of a factor of about 30×10^6. In fact, this factor is unlimited, since the communication is instantaneous and we can place the transmitter and receiver an arbitrarily long distance apart (although we may have to wait a while for the 'line' to be established).

Of course, if this scheme had any chance of working whatsoever, it would have been patented years ago. It does *not* work because an A photon passed through to the vertical polarizer is not automatically forced into a state of vertical polarization. According to quantum theory, it has equal probabilities for vertical or horizontal polarization, and we have no means of predicting in advance what the polarization will be. Thus, simply switching the A photon through to the vertical polarizer does not guarantee that photon B will be forced into a state of vertical polarization. In fact, despite switching between either path in the transmitter, there is still an unpredictable 50: 50 chance that photon B will be transmitted or blocked by the polarizer in the receiver. No message can be sent. (The AMAZING Company of Reading, U.K., recently filed for intellectual bankruptcy.)

Actually, it has been argued that our inability to exploit the apparent faster-than-light signalling between distant correlated photons allows quantum theory and special relativity peacefully to coexist. Special relativity is founded on the postulate that the speed of light represents the ultimate speed of transmission of any *conventional* signal. Whatever the nature of the communication between distant correlated photons, it is certainly not conventional.

4.5 DELAYED-CHOICE EXPERIMENTS

All our attention in this chapter has so far focused on the properties and behaviour of artificially generated correlated quantum particles. The experiments performed by Aspect and his colleagues were rather esoteric, involving a complicated apparatus and a somewhat complicated analysis. They seem to take the interested spectator a long way from what might appear to be the heart of the matter: wave–particle duality. After all, it was Bohr's insistence on the complementary nature of wave and particle properties that became one of the foundation stones of the Copenhagen interpretation. The Aspect experiments demonstrate in a round-about way that this complementarity creates a direct conflict between quantum theory and local reality. Is there a less round-about way of showing this?

In the last section, we saw that closing the last loophole through which local reality could be saved and the experimental results still explained involved switching between different analyser orientations while the emitted photons were in flight. The choice between the nature of the measurement was therefore delayed with respect to the transitions that created the photons in the first place. Is it possible to make this a delayed choice between measuring devices of a more fundamental nature?

For example, in our discussion in Section 2.6, we imagined the situation in which a single photon passes through a double slit apparatus to impinge on a piece of photographic film. We know that if we allow a sufficient number of photons individually to pass through the slits, one at a time, then an interference pattern will be built up. This observation suggests that the passage of each photon is governed by wave interference effects, so that it has a greater probability of being detected (producing a spot on the film) in the region of a bright fringe (see Fig. 1.3). It would seem that the photon literally passes through both slits simultaneously and interferes with itself. As we noted in our earlier discussion, the sceptical physicist who places a detector over one of the slits in order to show that the photon passes through one or the other does indeed prove his point—the photon is detected, or not detected, at one slit. But then the interference pattern can no longer be observed.

Advocates of local hidden variables could argue that the photon is somehow affected by the way we choose to set up our measuring device. It thus adopts a certain set of physical characteristics (hidden variables) if the apparatus is set up to show particle-like behaviour, and adopts a different set of characteristics if the apparatus is set up to show wave interference effects. However, if we can design an apparatus that allows us to choose between these totally different kinds of measuring device, we could delay our choice until the photon was (according to a local hidden variable theory) 'committed' to showing one type of behaviour. We suppose that the photon cannot change its 'mind' *after* it has passed through the slits, when it discovers what kind of measurement is being made.

Photons have it both ways

In 1978, the physicist John Wheeler proposed just such a delayed-choice experiment, which is in effect a modified version of the double slit apparatus described above. This experiment has recently been performed in the laboratories of two independent groups of researchers: Carroll O. Alley, Oleg G. Jakubowicz and William C. Wickes from the University of Maryland and T. Hellmuth, H. Walther and Arthur G. Zajonc from the University of Munich. Both groups used a similar experimental approach, a somewhat simplified version of which is described below.

The basic apparatus is shown schematically in Fig. 4.13. A pulse of light from a laser was passed through an optical device called a beam-splitter which, like a half-silvered mirror, transmits half the intensity of the incident light and reflects the other half. The split light beams followed two paths, indicated as A and B in Fig. 4.13. Fully reflecting mirrors were used to bring the two beams back into coincidence inside a triangular prism.

The recombined beams show wave interference effects. Viewed in terms of a wave picture, the relative phases of the waves (positions of the peaks and troughs) at the point where the beams recombine determines whether they show constructive interference (peak coincides with peak) or destructive interference (peak coincides with trough). The relative phases of the waves could be adjusted simply by changing the length of one of the paths. In Fig. 4.13, a 'phase-shifter' is shown in path A.

In fact, the dashed line drawn inside the triangular prism represents another beamsplitting surface, arranged to provide another 90° phase difference between light reflected from it and transmitted through it. Light reflected from this surface was detected by photomultiplier 1, and transmitted light was detected by photomultiplier 2. Thus, if the light waves entering the prism from paths A and B were already out of phase by 90° as a result of the different lengths of the paths, then the light reflected from the beamsplitting surface was 180° out of phase (peak

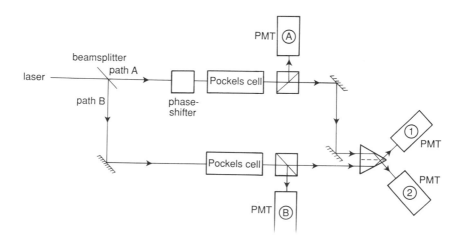

Fig. 4.13 Schematic diagram of the experimental apparatus used by Alley and colleagues to perform delayed-choice measurements.

coincident with trough), giving destructive interference. No light was detected by photomultiplier 1. On the other hand, light transmitted through the prism was shifted back into phase (peak coincident with peak), giving constructive interference. This light was detected by photomultiplier 2. The advantage of this arrangement was that interference effects were readily observed by the simple fact that all the light was detected by only one photomultiplier (photomultiplier 2). Blocking one of the paths, and thereby preventing the possibility of interference, resulted in equal light intensities reaching these photomultipliers.

Performing the experiment with the laser light intensity reduced, so that only one photon passed through the apparatus at a time, resulted in the expected detection of the photons only by photomultiplier 2. In this arrangement, the photon behaved as though it had passed along both paths simultaneously, interfering with itself inside the triangular prism, in exact analogy with the double slit experiment.

The researchers also inserted two optical devices called Pockels cells, one in each path. Without going into too many details, a Pockels cell consists of a crystal across which a small voltage is applied. The applied electric field induces birefringence in the crystal—in effect, it becomes a polarization rotator. Vertically polarized light passing through a birefringent crystal can, if the conditions are right, emerge horizontally polarized. A permanent polarizing filter was used in conjunction with each Pockels cell to reflect any horizontally polarized light out of the path and into a photomultiplier. Photomultiplier A monitored light reflected out of path A and photomultiplier B monitored light reflected out of path B.

If both Pockels cells were switched off (no voltage applied), the vertically polarized light passed down both paths undisturbed and recombined in the triangular prism to show interference effects. If either Pockels cell was switched on, the vertically polarized light passing through the active cell became horizontally polarized and was deflected out of its path and detected, preventing the observation of interference effects. Thus, with only one photon in the apparatus, switching on either Pockels cell was equivalent to asking which path through the apparatus the photon had taken. (For example, its detection by photomultiplier A showed that it had passed along path A.) This is analogous to asking which slit the photon goes through in the double slit experiment.

The choice between measuring a single photon's wave-like properties (passing along both paths) or particle-like properties (passing along one path only) was therefore made by switching on one of the Pockels cells. The great advantage of this arrangement was that this switching could be done within about 9 ns. The lengths of the paths A and B were each about 4.3 m, which a photon moving at the speed of light can cover in about 14.5 ns. Thus, the choice of measuring device could be made *after* the photon had interacted with the beamsplitter. There was therefore no way the photon could 'know' in advance whether it should pass along both paths to show wave interference effects (both Pockels cells off) or if it should pass along only one of the paths to show localized particle-like properties (one Pockels cell on).

Both groups of researchers reported results in agreement with the expectations of quantum theory. Within the limitations set by the instruments, with one of the Pockels cells on photons were indeed detected in one or other of the two paths and no interference could be observed. With both Pockels cells off, photons were detected only by photomultiplier 2, indicative of wave interference effects. Of course, according to the Copenhagen interpretation, the wavefunction of the photon develops along both paths. If one of the Pockels cells is switched on, the detection of a photon directed out of either path collapses the wavefunction instantaneously, and we infer that the photon was localized in one or other of the two paths.

Wheeler's 'Great Smoky Dragon'

John Wheeler has described this behaviour in a particularly picturesque way. Like the photon entering the delayed-choice apparatus, Wheeler's

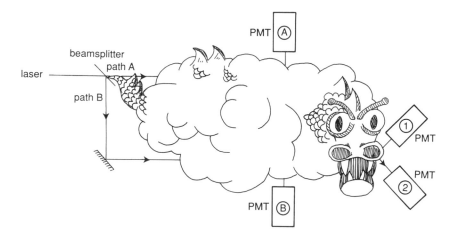

Fig. 4.14 Wheeler's 'Great Smoky Dragon'. Based on the drawing by Field Gilbert for John Wheeler which appears in French, A.P. and Kennedy, P.J. (eds) (1985). *Niels Bohr: a centenary volume*, Harvard University Press, Cambridge, MA, p. 151.

'Great Smoky Dragon', depicted in Fig. 4.14, has a sharply defined tail. Our knowledge of the tail therefore seems complete and unambiguous. The point at which the photon is detected — the mouth of the Dragon — is similarly sharp and clear to us. However, the middle of the Dragon is a fog of uncertainty: '. . . in between we have no right to speak about what is present.'

Wheeler has also suggested that the delayed-choice experiment can be performed on a cosmological scale, by making use of the gravitational lens effect. Two close-lying quasi-stellar objects (quasars), labelled 0957 + 561A, B are believed to be one and the same quasar. One image is formed by light emitted directly towards earth from the quasar. A second, virtual, image is produced by light emitted from the quasar which would normally pass by the earth but which is bent back by an intervening galaxy (this is the gravitational lens effect — see Fig. 4.15). The light reaching earth from the quasar can therefore travel by two paths. If we choose to combine the light from these paths we can, in principle, obtain interference effects.

We seem to have the power to decide by what route (or routes) any given photon emitted from the quasar travels to earth billions of years *after* it set out on its journey. Wheeler wrote:[†]

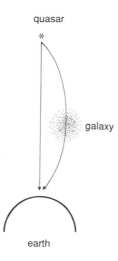

Fig. 4.15 The gravitational lens effect offers a means of performing the delayed-choice experiment on a cosmological scale.

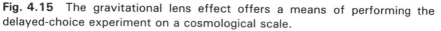

[†] Wheeler, J. A. (1981) in *The American Philosophical Society and The Royal Society: papers read at a meeting, June 5*. American Philosophical Society, Philadelphia. Reproduced in Wheeler, J. A. and Zurek, W. H. (eds.) (1983). *Quantum theory and measurement*. Princeton University Press.

. . . in a loose way of speaking, we decide what the photon *shall have done* after it has *already* done it. In actuality it is wrong to talk of the 'route' of the photon. For a proper way of speaking we recall once more that it makes no sense to talk of the phenomenon until it has been brought to a close by an irreversible act of amplification: 'No elementary phenomenon is a phenomenon until it is a registered (observed) phenomenon.'

4.6 RETROSPECTIVE

The great debate between Bohr and Einstein on the meaning of quantum theory centred around Einstein's realist philosophy: he was reluctant to abandon objective reality and strict causality. It has been argued that Einstein's insistence that quantum theory is incomplete did not automatically make him an advocate of hidden variables. However, as we have already discussed, it is difficult to imagine that the arguments put forward in the EPR paper suggest anything other than a version of quantum theory which explicitly allows for elements of local reality, i.e. a local hidden variable theory. There can therefore be no doubt that the experiments described in this chapter create insuperable difficulties for those who hold to Einstein's views.[†] And these are not the only experiments accessible with modern instrumentation — the last 10 years have seen a large number of experimental verifications of quantum interference effects which are most readily interpreted in terms of non-local interactions.

Was Einstein wrong?

It is certainly a strange irony of the history of science that Einstein, having laid the foundations of quantum theory through his revolutionary vision, should have become one of the theory's most determined critics. When he launched his attack on the theory in 1935 with his charge of incompleteness, he could not have possibly anticipated the work of Bell, 30 years later. At the time they were made, the arguments between Bohr and Einstein were purely academic arguments between two eminent

---

[†] There are some in the physics community who vehemently disagree with this statement. The 'insuperable' difficulties can be overcome, they argue, by dispensing with the central concept of the photon and returning to classical (and locally realist) electromagnetic wave fields supplemented by non-classical random fluctuations in the so-called zero-point field. These fluctuations are responsible for background 'noise', which the detectors in the Aspect experiments are set up specifically to discriminate against. The arguments in favour of this alternative theory were being developed and presented roughly at the time the original manuscript of this book was being drafted, and are not given in Chapter 5. The interested reader is therefore directed to Marshall, T. and Santos, E. (1988). *Foundations of Physics*, **18**, 185; Marshall, T. W. (1991). *ibid*, **21**, 209; Marshall, T. W. (1992). *ibid*, **22**, 363.

physicists with different personal philosophies, one positivist and one realist. Bell's theorem changed all that. The arguments became sharply focused on practical matters that could be put to the test in the laboratory. If, like the great majority of the physics community, we are prepared to accept that the Aspect experiments have been correctly interpreted, then we must also accept that Einstein's charge of incompleteness is unsubstantiated, at least in the spirit in which that charge was made in 1935.

How would Einstein have reacted to these results? Of course, any answer to such a question is bound to be subjective. However, from the glimpses of Einstein's thoughts and feelings which have been revealed in this book, it seems reasonable to suppose that he would have accepted the results (and their interpretation in terms of non-local behaviour) at face value. Not for him a relentless striving to find more loopholes through which local reality might be preserved. It is also reasonable to suppose that he would not have been persuaded by these results to change his position regarding the interpretation of quantum theory. While accepting that the results are correct, I suspect that he would have still maintained that their interpretation contains 'a certain unreasonableness'. He might have marvelled at the unexpected subtlety of nature, but his conviction that 'God does not play dice' was an unshakeable foundation on which he built his personal philosophy.

So, was Einstein wrong? In the sense that the EPR paper argued in favour of an objective reality for each quantum particle in a correlated pair independent of the other and of the measuring device, then the answer must be 'Yes'. But if we take a wider view and ask instead if Einstein was wrong to hold to the realist's belief that the physics of the universe should be objective and deterministic, then we must acknowledge that we cannot answer such a question. It is in the nature of theoretical science that there can be no such thing as certainty. A theory is only 'true' for as long as the majority of the scientific community maintain a consensus view that the theory is the one best able to explain the observations. And the story of quantum theory is not over yet.

Was Bohr right?

I feel sure that Bohr would have been delighted by the results of the experiments described in this chapter. They appear to be a powerful vindication of complementarity, and graphically demonstrate the central, crucial role of the measuring device. Perhaps Bohr would have been quick to point out that the methods used to predict the results of the complicated experiments on correlated pairs of quantum particles are actually based on some of the simplest of experimental observations with polarized light. Observations such as those which led to Malus's law formed the basis of our derivations of the projection amplitudes given in Table 2.2, and which were used in our analysis of the Aspect experiments in Section 4.4. Seen in this light, quantum theory is no more than a useful means of interrelating different experimental arrangements, allowing us to take the results from one to predict the outcome of another. We cannot go beyond this because, according to Bohr's positivist outlook, we have reached the limit of what is knowable. The questions we ask of nature must always be expressed in terms of some kind of macroscopic experimental arrangement.

Does this mean that the Aspect and delayed-choice experiments prove that the Copenhagen interpretation is the only possible interpretation of quantum theory? I do not think so. We should here recall the arguments made in Section 3.2: the Copenhagen interpretation insists that, in quantum theory, we have reached the limit of what we can know. Despite the fact that this interpretation emerges unscathed from the experimental tests described in this chapter, there are some physicists who argue that it offers nothing by way of explanation. The non-locality and indeterminism of the quantum world create tremendous difficulties of interpretation, which the Copenhagen view dismisses with a metaphorical shrug of the shoulders. For some physicists, this is not good enough. We will see in the next chapter that while some of the suggested alternative interpretations seem bizarre, they are in principle no less bizarre than the Copenhagen interpretation.

Readers inclined to a less metaphysical outlook might ponder the merits of such alternatives. Why bother to seek strange new theories when a much tried and tested theory is already available? Surely any alternative will be so contrived and artificial that it will be worthless compared with the simple elegance of quantum theory? But look once more at the postulates of quantum theory described in Section 2.2. What could be more contrived and artificial than the wavefunction? Where is the justification for postulate 1, apart from the fact that it yields a theory that works? What about the problems of quantum measurement highlighted by the paradox of Schrödinger's cat? If these questions cause you to stop and think, and perhaps reveal a hint of doubt in your mind, then you will see why some physicists continue to argue that the Copenhagen interpretation cannot be the answer. Bohr himself once said that: 'anyone who is not shocked by quantum theory has not understood it.'

5
What are the alternatives?

5.1 PILOT WAVES, POTENTIALS AND PROPENSITIES

So, where do we go from here? It is apparent from the last chapter that nature denies us the easy way out. We can rule out the idea of local hidden variables, one of the simpler solutions to the conceptual problems of quantum theory. Yet the Copenhagen interpretation is regarded by some to be no interpretation at all. Even those who adhere to the Copenhagen view tend to put aside or disregard the conceptual difficulties that it raises as they analyse the results from the latest particle accelerator experiment. This is not a very satisfactory situation.

In this final chapter, we will survey some of the alternatives to the Copenhagen interpretation that have been put forward in the years since quantum theory was first developed. Although these alternatives are quite different from one another and from the original theory, we will find that they possess a common thread. In every case, they attempt to avoid the conceptual problems by introducing some additional feature into the theory. This is at least consistent with Einstein's belief that quantum theory is somehow incomplete. Such features are designed either to bring back determinism and causality, or to break the infinite regress of the quantum measurement process as illustrated by the paradox of Schrödinger's cat. The fact that rational scientists are prepared to go to such lengths to obtain an aesthetically or metaphysically more appealing version of the theory demonstrates the extent of the discomfort they experience with the dogma of the Copenhagen school.

Of course, if they are to work effectively, none of these alternatives should make predictions which differ from those of orthodox quantum theory for any experiment yet performed. Few, if any, even hint at the possibility that they could be subjected to stringent test through experiment. For many scientists, who have been brought up to regard observation and experiment as the keys to unlocking the mysteries of the physical world, a theory that cannot be tested is of no practical value. However, we should perhaps recall that our ability to perform precise measurements on the world is a relatively new development in the history of man's search for understanding. Without the kind of speculative thinking that the positivists dismiss as non-scientific, there could have been no science (and, for that matter, no positivism) in the first place.

De Broglie's pilot waves

Look back at the description of the double slit experiment in Chapter 1 and, in particular, the results of the electron interference experiment shown in Fig. 1.3. Perhaps there is something about these results that seems to nag in the backs of our minds. Wave–particle duality is manifested by the appearance of bright spots on the photographic film, showing where individual particles have been detected, but grouped into alternate bright and dark bands characteristic of wave interference.

But wait a moment. It is clear that we can only detect particles, whether through the chemical processes occurring in a photographic emulsion, or through the physical processes occurring in a photomultiplier or similar device. We understand that this is so because these detection processes require that the initial interaction between object and measuring device involves a whole quantum particle which cannot be sub-divided. Thus, a single electron or photon interacts with an ion in the photographic emulsion, initiating a chain of chemical reactions which ultimately results in the precipitation of a large number of silver atoms. That initial interaction appears to localize the particle: it reacts with this particular ion at this particular place on the film.

The evidence for the quantum particle's wave-like properties derives from the pattern in which a large number of individual particles are detected. According to the Copenhagen interpretation, this pattern arises because the wavefunction or state vector of each quantum particle has greater amplitude in some regions of the film compared with others, owing to interference effects generated by its passage through the two slits. Before it is detected, a quantum particle is 'everywhere' on the film, but it has a greater probability of interacting with an ion in those regions of the film where $|\psi|^2$ is large. These regions become bright fringes.

As it appears that we can only detect particles, Einstein's suggestion that the particles are real entities that follow precisely defined trajectories is very persuasive. In Chapter 4 we dismissed the possibility that any such trajectories are determined by local hidden variables, but are there other ways in which the particles' motions might be predetermined?

In 1926, Louis de Broglie proposed an alternative to Born's probabilistic interpretation of the wavefunction. Suppose, he said, that quantum particles like electrons and photons are independently real particles, moving in a real field. This is different again from Schrödinger's wave mechanics, which attempted to explain everything in terms of waves only. De Broglie suggested that the equations of quantum mechanics admit a double solution: a continuous wave field which has a statistical significance and a point-like solution corresponding to a localized particle. The continuous wave field can be diffracted and can exhibit interference effects. The motion of a real particle is somehow tied to the wave field, so that it is more likely to follow a path in which the amplitude of the wave field is large. Thus, the square of the amplitude of the wave field is still related to the probability of 'finding' the particle, but this is now because the real particle, which is always localized, has a preference for regions of space in which the wave amplitude is large.

In terms of the double slit experiment, we can imagine that the wave field interferes with itself as it passes through the slits, producing a pattern of bands of alternating large and small amplitudes on the photographic film. As a particle moves in the field, it is *guided* by the field amplitude, and therefore has a greater probability of arriving at the film in a region which we will recognize as a bright fringe when a sufficient number of particles has been detected. The particle is not prevented from following a trajectory which leads to it being detected in the region of a dark fringe, but this is much less probable because the amplitude of the field along such a path is small.

In de Broglie's theory, the wave field acts as a pilot field, dictating the direction of motion of the particle according to wave interference effects. Unlike complementarity, which offers us a choice between waves *or* particles depending on the nature of the measuring device, de Broglie's pilot wave interpretation suggests that reality is composed of waves *and* particles.

De Broglie completed his theory early in 1927. At the fifth Solvay Conference in October that year, Einstein commented that he thought de Broglie was searching in the right direction. Remember it was Einstein's remark connecting the wavefunction with a 'ghost field' that had led Born to develop his probabilistic interpretation. However, de Broglie's proposal that such a field is physically real differs completely from Born's view that the wavefunction in some way represents our state of knowledge of the quantum particle.

But de Broglie's discussions with members of the Copenhagen school (notably Pauli) began to raise doubts in his own mind about the validity of his theory. Pauli criticised the pilot wave idea, giving much the same reasons that had ultimately led to the rejection of Schrödinger's wave field. By early 1928, de Broglie was beginning to have second thoughts about his theory, and did not include it in a course on wave mechanics he taught at the Faculté des Sciences in Paris later that year. In fact, de Broglie became a convert to the Copenhagen view.

It is important to note that the pilot wave theory is a hidden variable theory. The hidden variable is not the pilot wave itself — that is already adequately revealed in the properties and behaviour of the wavefunction of quantum theory. It is actually the particle position that is hidden. Now we know from the results of the experiments described in the last chapter that two correlated quantum particles cannot be locally real, and so the pilot wave idea can be sustained only if we acknowledge that influences between the two particles can be communicated at speeds faster than that of light. It seems that we cannot have it both ways: either quantum theory is already complete or we must introduce non-local hidden variables which, in turn, appear to make the theory incompatible with special relativity. Either way, it is very doubtful that Einstein would have been satisfied.

Quantum potentials

We saw in Section 4.1 that the American physicist David Bohm, initially an advocate of Bohr's complementarity idea, eventually became dissatisfied with the Copenhagen view. Strongly encouraged and influenced by Einstein, he sparked off a renewal of interest in the question of hidden variables through the two papers he published on this subject in 1952 in the journal *Physical Review*. Bohm's hidden variable theory has much in common with de Broglie's pilot wave idea. However, Bohm continued to develop and refine his theory, despite the general indifference of the majority of the physics community.

The development of Bohm's theory involves a reworking or reinterpretation of the wavefunction as representing an objectively real field. To every real particle in this field, Bohm ascribed a precisely defined position *and* a momentum. He simply assumed that the wavefunction of the field can be written in the form $\psi = Re^{iS/\hbar}$, where R is an amplitude function and S is a phase function. Of course, $|\psi|^2 = |R|^2$ and so the probability of 'finding' the particle at a particular position is related to the modulus-squared of the amplitude function, as in orthodox quantum theory. In Bohm's theory, the average particle momentum is related to the phase function. The resulting laws of motion of the particle are governed by the usual classical potential energy V, and an additional *quantum potential* U which depends on the amplitude function: $U = (-\hbar^2/2m)\, \nabla^2 R / R$.

Most importantly, Bohm found that the quantum potential depends only on the *mathematical form* of the wavefunction, not its amplitude. Thus, the effect of the quantum potential can be large even in regions of space where the amplitude of the wavefunction is small. This contrasts with the effects exerted by classical potentials (such as a Newtonian gravitational potential), which tend to fall off with distance. A particle moving in a region of space in which no classical potential is present can therefore still be influenced by the quantum potential. As before, the wavefunction has a dual role – it is the function from which probabilities can be obtained in the usual way and it can also be used to derive the shape of the quantum potential.

For example, when the field passes through a double slit apparatus, the resulting interference effects generate a complicated quantum potential. Theoretical calculations of the shape of this potential have been done for the case of a single electron passing through the apparatus, and the results of these are shown in Fig. 5.1. The possible trajectories of the electron through either of the slits are determined by the quantum potential. Figure 5.2 shows theoretical trajectories corresponding

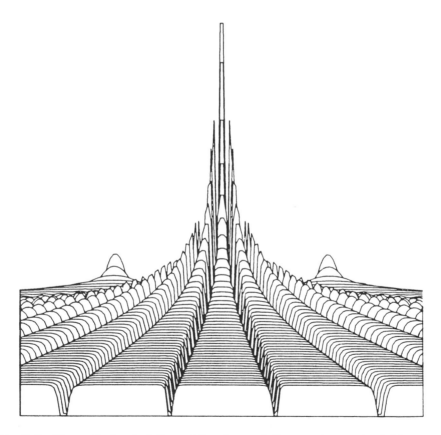

Fig. 5.1 Theoretical calculation of the shape of the quantum potential for an electron passing through a double slit apparatus. Reprinted with permission from J. P. Vigier *et al.* (1987). *Quantum implications*, (ed. B. J. Hiley and F. D. Peat). Routledge and Kegan Paul, London.

slit A   slit B

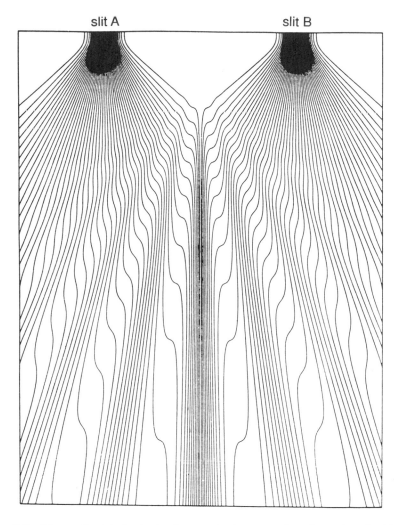

Fig. 5.2 Theoretical trajectories for an electron passing through a double slit apparatus, calculated using the quantum potential shown in Fig. 5.1. Reprinted with permission from J. P. Vigier *et al.* (1987). *Quantum implications*, (ed. B. J. Hiley and F. D. Peat). Routledge and Kegan Paul, London.

to the potential shown in Fig. 5.1. Note how these trajectories group together to produce (after the detection of many electrons) a set of alternating bright and dark fringes.

The quantum potential is the medium through which influences on distant parts of a correlated quantum system are transmitted. The measurement of some property (such as vertical polarization) of one of a pair of correlated photons instantaneously changes the quantum potential in a non-local manner, so that the other particle takes on the required properties without the need for a collapse of the wavefunction. We saw in Section 4.4 that such an instantaneous transmission cannot be exploited to send coded information, and so conflict with the postulates of special relativity might in principle be avoided. Although such influences are transmitted at speeds faster than that of light, they represent entirely causal connections between the particles. Furthermore, there is absolutely no conflict here with Bohr's contention that the measuring device has a fundamental role which cannot be ignored. In Bohr's theory, changing the measuring device (which might amount to no more than changing the orientation of a polarizing filter) instantaneously changes the wavefunction and hence the quantum potential: all future trajectories of quantum particles passing through the apparatus are thus predetermined. The quantum potential effectively interconnects every region of space into an inseparable whole.

This aspect of 'wholeness' is central to Bohm's theory, as indeed it is to Bohr's. Our day-to-day use of the quantum theory depends on our ability to factorize the wavefunction into more manageable parts (for example, in an approximation routinely applied in chemical spectroscopy, a molecular wavefunction is factorized into separate electronic, vibrational, rotational and translational parts). Under some circumstances, the wavefunction, and hence the quantum potential, can be factorized into a discrete set of sub-units of the whole. However, when we come to deal with experiments on pairs of correlated quantum particles, we should not be surprised if the wavefunction cannot be factorized in this way. The non-local connections between distant parts of a quantum system are determined by the wavefunction of the whole system. In one sense, Bohm's theory takes a 'top-down' approach: the whole has much greater significance the sum of its parts and, indeed, determine the behaviour and properties of its parts. Contrast this with the 'bottom-up' approach of classical physics, in which the behaviour and properties of the parts determines the behaviour of the whole.

De Broglie himself initially rejected Bohm's theory, for the same reasons that he had abandoned his own pilot wave approach more than 20 years earlier. However, he eventually came to be persuaded that some of the problems raised by identifying the wavefunction as a real field could, in fact, be overcome.

The implicate order

During the 1960s and 1970s, Bohm delved more deeply into the whole question of *order* in the universe. He developed a new approach to understanding the quantum world and its relationship with the classical world which contains, and yet transcends, Bohr's notion of complementarity. Bohm has described one early influence as follows:[†]

> . . . I saw a programme on BBC television showing a device in which an ink drop was spread out through a cylinder of glycerine and then brought back together again, to be reconstituted essentially as it was before. This immediately struck me as very relevant to the question of order, since, when the ink drop was spread out, it still had a 'hidden' (i.e. non-manifest) order that was revealed when it was reconstituted. On the other hand, in our usual language, we would say that the ink was in a state of 'disorder' when it was diffused through the glycerine. This led me to see that new notions of order must be involved here.

Bohm reasoned that the order (the localized ink drop) becomes *enfolded* as it is diffused through the glycerine. However, the information content of the system is not lost as a result of this enfoldment: the order simply becomes an implicit or *implicate* order. The ink drop is reconstituted in a process of *unfoldment*, in which the implicate order becomes, once again, an explicate order that we can readily perceive.

Bohm went further in his book *Wholeness and the implicate order*. He recognized that the equations of quantum theory describe a similar enfoldment and unfoldment of the wavefunction. We understand and interpret classical physics in terms of the behaviour of material particles moving through space. The order of the classical world is therefore enfolded and unfolded through this fundamental motion. In Bohm's quantum world the acts of enfoldment and unfoldment are themselves fundamental. Thus, all the features of the physical world which we can perceive and which we can subject to experiment (the explicate order) are realizations of potentialities contained in the implicate order. The implicate order not only contains these potentialities but also determines which will be realized. Bohm has written:[†] '. . . the implicate order provided an image, a kind of metaphor, for intuitively understanding the implication of wholeness which is the most important new feature of the quantum theory'. With one very important exception, which we will examine later in this chapter, the implicate order represents a kind of ultimate hidden variable—a deeper reality which is revealed to us through the unfoldment of the wavefunction.

Bohm has extended and adapted his original hidden variable theory, guided by the holistic approach afforded by his theory of the implicate order. By modifying the equations of quantum field theory, he has done

---

† Bohm, D., in Hiley, B.J. and Peat, F.D. (eds.) (1987). *Quantum implications*. Routledge and Kegan Paul, London.

away with the need to invoke the existence of independent, objectively real particles. Instead, particle-like behaviour results from the convergence of waves at particular points in space. The waves repeatedly spread out and reconverge, producing 'average' particle-like properties, corresponding to the constant enfoldment and unfoldment of the wavefunction. This 'breathing' motion is governed by a super quantum potential, related to the wavefunction of the whole universe. 'We have a universal process of constant creation and annihilation, determined through the super quantum potential so as to give rise to a world of form and structure in which all manifest features are only relatively constant, recurrent and stable aspects of this whole.'[†]

Pure metaphysics? Certainly. But Bohm has done nothing more than adopt a particular philosophical position in deriving his own cosmology. As we have seen, analysis of the Copenhagen interpretation reveals that it too is really nothing more than a different philosophical position. The difference between these two is that the philosophy of the Copenhagen school is made 'scientific' through the use of the (entirely arbitrary) postulates of quantum theory. Bohm has argued that the reason orthodox quantum theory is derived from these postulates rather than postulates based on an implicate order or similar construction is merely a matter of historical precedent.

Popper's propensities

Karl Popper is one of this century's most influential philosophers of science. Born in Vienna in 1902, he discussed the interpretation of quantum theory directly with its founders: Einstein, Bohr, Schrödinger, Heisenberg, Pauli, *et al.* At 89, he actively continues the debate, adding to a prolific output of writings on the subjects of quantum theory, the philosophy of science and the evolution of knowledge. This output began in 1934 with the publication, in Vienna, of his now famous book *Logik der forschung*, first published in English in 1959 as *The logic of scientific discovery*. The basic tenets of Popper's philosophy—particularly his principle of falsifiability—will be familiar to anyone who has delved (even superficially) into the philosophy of science.

Popper's position on quantum theory is easily summarized: he is a realist. While not agreeing in total with all the ideas advanced by Einstein and Schrödinger, it is clear from his writings that he stands in direct opposition to the Copenhagen interpretation, and in particular to the positivism of the young Heisenberg. Although Popper interacted with

---

[†] Bohm, D., in Hiley, B. J. and Peat, F. D. (eds.) (1987). *Quantum implications.* Routledge and Kegan Paul, London.

various members of the Vienna Circle (particularly Rudolph Carnap) he did not share the Circle's philosophical outlook. Inspired instead by the Polish philosopher Alfred Tarski, Popper was motivated by a desire to search for *objective* truth, a motivation that he held in common with Carnap although their methods differed considerably.

During this century there has been an important debate between philosophers and scientists concerning the nature of probability. In 1959, Popper published details of his own *propensity* interpretation of probability which has implications for quantum probabilities. This interpretation is best illustrated by reference to a simple example, and we will use here an example used extensively by Popper himself.

The grid shown in Fig. 5.3 represents an array of metal pins embedded in a wooden board. One end of this pinboard is raised so as to make a slight incline. A small marble, selected so that it just fits between any two adjacent pins, is rolled down the board and enters the grid at its centre, as shown. On striking a pin, the marble may move either to the left or to the right. The path followed by the marble is then determined by the sequence of random left versus right jumps as it hits successive pins. We measure the position at the bottom of the grid at which the marble exits.

Repeated measurements made with one marble (or with a 'beam' of identical marbles) allow us to determine the frequencies with which the individual marbles exit at specific places on the grid. These we can turn into statistical probabilities in the usual way (see Section 3.5). If the

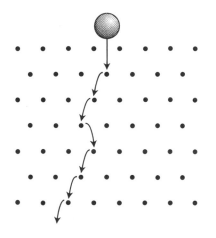

Fig. 5.3 Popper's pinboard. The dots represent pins in a pinboard through which a marble passes in a sequence of left or right jumps. The propensity for the marble to exit the grid at a particular point is determined by the properties of the marble *and* the grid as a whole.

marble(s) always enter the grid at the same point and if the pins are identical, then we would expect a uniform distribution of probabilities, with a maximum around the centre and thinning out towards the extreme left and right. The shape of the distribution simply reflects the fact that the probability of a sequence of jumps in which there are about as many left jumps as there are right is greater than the probability of obtaining a sequence in which the marble jumps predominantly to the left or right.

Popper has argued that each probability is determined by the *propensity* of the system as a whole to produce a specific result. This propensity is a property of the marble *and* the 'apparatus' (the pinboard). Change the apparatus, perhaps by removing one of the pins, and the propensities of the system — and hence the probabilities of obtaining specific results — change instantaneously even though the paths of individual marbles may not take them anywhere near the region of the missing pin.

According to Popper, reality is composed of particles only. The wave-function of quantum theory is a purely statistical function, representing the propensities of the particles to produce particular results for a particular experimental arrangement. Change the arrangement (by changing the orientation of a polarizing filter or by closing a slit) and the propensities of the system — and hence the probability distribution or wavefunction — changes instantaneously. For him, the Heisenberg uncertainty relations are merely relations representing the scattering of objectively real particles.

Popper's interpretation does have an intuitive appeal. The collapse of the wavefunction does not represent a physical change in the quantum system but rather a change in the state of our knowledge of it. Going back to the pinboard, before the 'measurement', the marble can exit from the grid at any position and the probabilities for each are determined by the propensities inherent in the system. During the measurement, the marble is observed to exit from one position only. Of course, the probabilities themselves have not changed, as is readily shown by repeating the measurement with another marble, but the *system* has changed. We can define a new set of probabilities for the new system: the probability for the marble to be found at its observed point of exit being unity and all others being zero.

This last point can be made clear with the aid of another example drawn from the quantum world. Imagine a photon impinging on a half-silvered mirror. Suppose that the probability that the photon is transmitted through the mirror is equal to the probability that it is reflected and, for simplicity, we set these equal to $\frac{1}{2}$. These probabilities are related to the propensities for the system (photon + mirror and detectors). We can write

$$p(a, b) = p(-a, b) = \frac{1}{2},$$ (5.1)

where $p(a, b)$ is the probability of detecting a transmitted photon a relative to the system before the measurement b, and $p(-a, b)$ is the probability of detecting a reflected photon, 'not a' or $-a$, relative to the system b. Now suppose that we detect a transmitted photon. According to Popper, the system (photon + mirror and detectors) has now completely changed and it is necessary to define two new probabilities relative to the new system. Because a transmitted photon has been detected, these probabilities are

$$p(a, a) = 1 \text{ and } p(-a, a) = 0.$$ (5.2)

The original probabilities $p(a, b)$ and $p(-a, b)$ have not changed since they refer to the system *before* the measurement was made. These probabilities apply whenever the experiment is repeated.

It is only in what Popper calls the 'great quantum muddle' that $p(a, a)$ is identified with $p(a, b)$ and $p(-a, a)$ with $p(-a, b)$ and the process referred to as the collapse of the wavefunction. He writes:[†]

No *action* is exerted upon the [wavefunction], neither an action at a distance nor any other action. For $p(a, b)$ is the propensity of the state of the photon relative to the original experimental conditions . . . the reduction of the [wavefunction] clearly has nothing to do with quantum theory: it is a trivial feature of probability theory that, whatever a may be, $p(a, a) = 1$ and (in general) $p(-a, a) = 0$.

However, the propensity interpretation runs into some difficulties when we attempt to use it to explain the wave-like behaviour of quantum particles. In particular, the only way to explain wave interference effects is to suggest, as Popper does, that the propensities themselves can somehow interfere. Popper concludes that this interference is evidence that the propensities are physically real rather than simply mathematical devices used to relate the experimental arrangement to a set of probabilities. He thus writes of particles and their associated 'propensity waves' or 'propensity fields'. This is clearly taking us back towards de Broglie's pilot wave idea and, indeed, Popper has noted that:[†] 'As to the pilot waves of de Broglie, they can, I suggest, be best interpreted as waves of propensities.'

As we explained above, the pilot wave theory is a hidden variable theory and we have previously come to the conclusion that no local hidden variable theory can account for the results of the Aspect experi-

---

[†] Popper, K. R. (1982). *Quantum theory and the schism in physics*. Unwin Hyman, London.

ments. We saw that David Bohm had earlier decided not to be limited by the constraints imposed by the postulates of special relativity in developing his own non-local version of the theory. Initially, Popper rebelled against taking this step, agreeing with Einstein that the idea of superluminal influences passing between two distant correlated quantum particles 'has nothing to recommend it'. However, Popper's views changed as the experimental results became increasingly difficult to explain in terms of any locally real theory. If it is accepted that there can be non-local, superluminal influences transmitted via the propensity field, then there appears to be little to choose between Popper's approach and Bohm's idea of the implicate order.

5.2 AN IRREVERSIBLE ACT

Perhaps the greatest source of discomfort that scientists experience with the Copenhagen interpretation of quantum theory arises from its treatment of quantum measurement. As we pointed out in Section 2.6, given some initial set of conditions, the equations of quantum theory describe the future time evolution of a wavefunction or state vector in a way which is quite deterministic. The wavefunction moves through Hilbert space in a manner completely analogous to a classical wave moving through Euclidean space. If we are able to calculate a map of the amplitude of the wavefunction in Hilbert space, we can use quantum theory to tell us what this map should look like at some later time.

However, when we come to consider a measurement, then the Copenhagen interpretation requires us to set aside these elegant deterministic equations and reach for a completely different tool. These equations do not allow us to compute the probabilities for the wavefunction to be projected into one of a set of measurement eigenfunctions: this must be done in a separate step. The measurement eigenfunctions are determined at the whim of the observer, but which result will be obtained with any one quantum particle is quite indeterminate. And we learn from Schrödinger's cat that quantum theory has nothing whatsoever to say about where in the measurement process this projection or collapse of the wavefunction takes place.

It is true that most scientists are primarily concerned about the deterministic part of quantum theory in that they are interested in using it to picture how atoms or molecules behave in the absence of an interfering observer. For example, molecular quantum theory can provide beautiful pictures of molecular electronic orbitals which we can use to understand chemical structure, bonding and spectroscopy. Little thought is given to what these pictures might mean in the context of a measurement — it is enough for us to use them to imagine how molecules are, independently of ourselves and our instruments. But our information is derived from measurements. It is derived from processes in which the nice deterministic equations of motion do not apply. It is derived from processes which present us with profound conceptual difficulties. The search for solutions to the quantum measurement problem has produced some spectacularly bizarre suggestions. We will consider some of these here and in the next two sections.

The arrow of time

It is our general experience that, apart from in a few science fiction novels, time flows only one way: forwards. Why? The equations of both classical and quantum mechanics appear quite indifferent to the direction in which time flows. With the possible exception of the collapse of the wavefunction (which we will discuss at length below), replacing t by $-t$ in the equations of classical or quantum mechanics makes no difference to the validity or applicability of the equations. When we abandon the idea of an absolute time, as special relativity demands, the equations do not even recognize a 'now' distinguishable from the past or

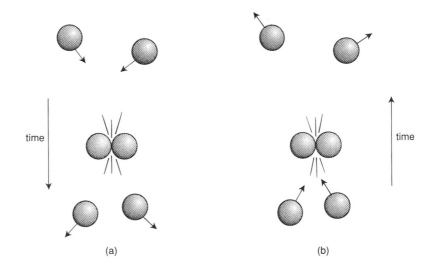

Fig. 5.4 (a) A collision between two atoms (pictured here as rigid spheres) seen in forward time. (b) The time-reversal of (a), in which the momenta of the atoms are exactly reversed. The collision in (b) looks no more unusual than that in (a).

future. But our perceptions are quite different: the flow of time is an extremely important part of our conscious existence.

Imagine a collision between two atoms (Fig. 5.4). The atoms come together, collide with each other and move apart in different directions with different velocities. Run this picture backwards in time and we see nothing out of the ordinary: the atoms come together, collide and move apart. The physics of the exact time-reverse of the collision is no different in principle from the physics viewed in forward time.

Now imagine a collision between an atom and a diatomic molecule (Fig. 5.5). This time, we suppose that the collision is so violent that it smashes the molecule into two atomic fragments. All three atoms move apart in directions and with velocities which are themselves determined by the initial conditions. Again, the equations are indifferent to this collision run in reverse: bring together the three atoms with exactly the opposite momenta and the molecule will re-form. However, we now sense that this process looks 'wrong' when run in reverse, or at least looks very unlikely.

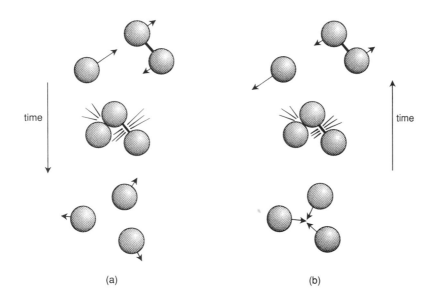

(a)   (b)

Fig. 5.5 (a) A collision between an atom and a diatomic molecule in which the molecule is dissociated into two atoms, seen in forward time. (b) The time-reversal of (a), in which the three atoms come together and a diatomic molecule forms. Now the time-reversed collision looks 'odd' in the sense that it seems a most unlikely event.

This picture looks wrong because it is our general experience that a system does not spontaneously transform from a more complicated to a less complicated state (a broken glass spontaneously reassembling itself, for example). The important difference between the time-reversed collisions shown in Figs. 5.4 and 5.5 is that in Fig. 5.5, the number of degrees of freedom is larger — we need more position and velocity coordinates to describe (mathematically) what is going on. This tendency for the physical world always to transform (or disperse) into something more complicated is embodied in the second law of thermodynamics, which can be stated as follows:

For a spontaneous change, the entropy of an isolated system always increases.

Students of science are usually taught to understand entropy as a measure of the 'disorder' in a system. Thus, a crystal lattice has a very ordered structure, with its constituent atoms or molecules arranged in a regular array, and it therefore has a low entropy. On the other hand, a gas consists of atoms or molecules moving randomly through space, colliding with each other and with the walls of the container, and therefore has a much higher entropy. This is confirmed by experimental thermodynamics: diamond has an entropy $S°$ of $2.4 \, \text{J} \, \text{K}^{-1} \, \text{mol}^{-1}$ at 298 K and 1 bar pressure. This figure should be compared with the entropy of gaseous carbon atoms, $S° = 158 \, \text{J} \, \text{K}^{-1} \, \text{mol}^{-1}$.

The second law of thermodynamics refers to spontaneous or *irreversible* changes. For a reversible change — one in which we keep track of all the motions in the system and can at any time apply (in principle) an infinitesimal force to reverse their directions — the entropy does not increase, but can be moved from one part of the system to another. The most important aspect of the second law is that it appears to embody a unidirectional time. All spontaneous changes taking place in an isolated system increase the entropy as time increases. We cannot decrease the entropy without directly interfering with the system (for example, restoring a broken glass to its former state). But then the system is no longer isolated, and when we come to consider the larger system and take account of the methods used to re-melt the glass and re-form its original shape, we find that the entropy of this larger system will have increased. A spontaneous change in which the entropy of an isolated system decreased would effectively be running backwards in time.

Just exactly where does this time asymmetry come from? It is not there in the classical equations of motion, and yet it is such an obvious and important part of our experience of the world. The second law is really a summary of this experience.

With the emergence of Boltzmann's statistical mechanics, a new comprehension of entropy as a measure of probability became possible. On the molecular level, we can now understand the second law in terms of the spontaneous transition of a system from a less probable to a more probable state. A gas expands into a vacuum and evolves in time towards the most probable state in which the density of its constituent atoms or molecules is uniform. We call this most probable state the equilibrium state of the gas. However, this talk of probabilities introduces a rather interesting possibility. A spontaneous transition from a more probable to a less probable state (decreasing entropy) is not disallowed by statistical mechanics — it merely has a very low probability of occurring. Thus, the spontaneous aggregation of all the air molecules into one corner of a room is not impossible, just very improbable. The theory seems to suggest that if we wait long enough (admittedly, much longer than the present age of the universe), such improbable spontaneous entropy-reducing changes will eventually occur. Some scientists (including Einstein) concluded from this that irreversible change is an illusion: an apparently irreversible process will be reversed if we have the patience to wait. We will return to this argument below.

If spontaneous change must always be associated with increasing disorder, how do we explain the highly ordered structures (such as galaxies and living things) that have evolved in the universe? Some answers are being supplied by the new theory of *chaos*, which describes how amazingly ordered structures can be formed in systems far from equilibrium.

Time asymmetry and quantum measurement

What does all this have to do with quantum measurement? Well, quite a lot actually. However, to see how arguments about spontaneous changes and the second law fit into the picture, it is necessary to step beyond the boundaries of undergraduate physics and chemistry and delve a little into quantum statistical mechanics. It is neither desirable nor really necessary for us to go too deeply into this subject in this book. Instead, we will draw on some useful concepts, basic observations and ideas that have been presented in greater detail elsewhere (see the bibliography for some excellent references on this subject).

Quantum statistical mechanics is essentially a statistical theory concerned with collections (or, more correctly, ensembles) of quantum particles. Consider an ensemble of N quantum particles all present in a quantum state denoted $|\Psi\rangle$. Such an ensemble is said to be in a *pure state*. The state vector of each particle in the ensemble can be expressed as a superposition of the eigenstates of the operator corresponding to some measuring device. Suppose there are n of these eigenstates; $|\psi_1\rangle$, $|\psi_2\rangle$, $|\psi_3\rangle$, . . ., $|\psi_n\rangle$. We know from our discussion in

Chapter 2 that the probability for any particle in the ensemble to be projected into a particular measurement eigenstate is given by the modulus-squared of the corresponding projection amplitude (or the coefficient in the expansion). After the measurement has taken place, each particle in the ensemble will have been projected into one, and only one, of the possible measurement eigenstates. The quantum state of the ensemble is now a *mixture*, the number of particles present in a particular eigenstate being proportional to the modulus-squares of the projection amplitudes. Pictorially, the process can be written thus:

$$
\left\{ |\Psi\rangle \right\} \quad \xrightarrow[\text{measurement}]{\text{quantum}} \quad
\left\{
\begin{array}{l}
|\psi_1\rangle \\
|\psi_2\rangle \\
|\psi_3\rangle \\
\cdot \\
\cdot \\
\cdot \\
|\psi_n\rangle
\end{array}
\right\}
\quad
\begin{array}{l}
N|\langle \psi_1 | \Psi \rangle|^2 \text{ particles} \\
N|\langle \psi_2 | \Psi \rangle|^2 \text{ particles} \\
N|\langle \psi_3 | \Psi \rangle|^2 \text{ particles} \\
\cdot \\
\cdot \\
\cdot \\
N|\langle \psi_n | \Psi \rangle|^2 \text{ particles}
\end{array}
$$

ensemble of N particles

Pure state   Mixture

This is just another way of looking at the problem of the collapse of the wavefunction. The act of quantum measurement transforms a pure state into a mixture. The mathematician John von Neumann showed that this transformation is associated in quantum statistical mechanics with an increase in entropy. Thus, irreversibility or time asymmetry appears as an intrinsic feature of quantum measurement.

The problem now is that the equations of motion derived from the time-dependent Schrödinger equation do not allow such a transformation. As we described in Section 2.6, if a quantum system starts as a pure state, it will evolve in time as a pure state according to the equations of motion. This is because, in mathematical terms, the action of the time evolution operator in transforming a wavefunction at some time t into the same wavefunction at some later time t' is equivalent in many ways to a simple change of coordinates. Abrupt, irreversible transformation into a mixture of states is possible only in quantum measurement through the collapse of the wavefunction.

From being to becoming

The Nobel prize-winning physical chemist Ilya Prigogine has argued that we are dealing here with two different types of physics. He identifies a physics of *being*, associated with the reversible, time-symmetric equations of classical and quantum mechanics, and a physics of *becoming*, associated with irreversible, time-asymmetric processes which increase the entropy of an isolated system. He rejects the argument that irreversibility is an illusion or approximation introduced by us, the observers, on a completely reversible world. Instead, he advocates a 'new complementarity' between dynamical (time-symmetric) and thermodynamic (time-asymmetric) descriptions. This he does in an entirely formal way by defining an explicit microscopic *operator* for entropy and showing that it does not commute with the operator governing the time-symmetric dynamical evolution of a quantum system.

According to Prigogine, introducing a microscopic entropy operator has certain consequences for the equations describing the dynamics of quantum systems. Specifically, he shows that the equations now consist of two parts—a reversible, time-symmetric part equivalent to the usual description of quantum state dynamics and a new irreversible, time-asymmetric part equivalent to an 'entropy generator'. Prigogine's approach is *not* to attempt to derive the second law from the dynamics of quantum particles but to assume its validity and then seek ways to introduce it *alongside* the dynamics. In his book *From being to becoming*, published in 1980, he wrote:[†]

The classical order was: particles first, the second law later—being before becoming! It is possible that this is no longer so when we come to the level of elementary particles and that here we must *first* introduce the second law before being able to define the entities.

It is interesting to note that Prigogine's approach parallels that of Boltzmann a century earlier. Boltzmann attempted to find a molecular mechanism that would ensure that a non-equilibrium distribution of molecular velocities in a gas would evolve in time to a Maxwell (equilibrium) distribution. The result was a dynamical equation that contains both reversible and irreversible parts, the latter providing an entropy increase independently of the exact nature of the interactions between the molecules. Like Prigogine, Boltzmann could not derive this equation from classical dynamics—he just had to assume it.

Prigogine concludes his book with the observation that:[†]

The basis of the vision of classical physics was the conviction that the future is determined by the present, and therefore a careful study of the present permits the unveiling of the future. At no time, however, was this more than a theoretical possibility.

Indeed, one of the most important lessons to be learned from the new

---

[†] Prigogine, Ilya (1980). *From being to becoming.* W. H. Freeman, San Francisco, CA.

theory of chaos is that, even in classical mechanics, our ability to predict the future behaviour of a dynamical system depends crucially on our knowing exactly its initial conditions. The smallest differences between one set of initial conditions and another can lead to very large differences in the subsequent behaviour, and it is becoming increasingly apparent that in complex systems we simply cannot know the initial conditions precisely enough. This is not because of any technical limitation on our ability to determine the initial conditions, it is a reflection of the fact that predicting the future would require *infinitely* precise knowledge of these conditions.

Prigogine again:[†]

Theoretical reversibility arises from the use of idealisations in classical or quantum mechanics that go beyond the possibilities of measurement performed with any finite precision. The irreversibility that we observe is a feature of theories that take proper account of the nature and limitation of observation.

In other words, it is reversibility, not irreversibility, which is an illusion: a construction we use to simplify theoretical physics and chemistry.

A bridge between worlds

Bohr recognized the importance of the 'irreversible act' of measurement linking the macroscopic world of measuring devices and the microscopic world of quantum particles. Some years later, John Wheeler wrote about an 'irreversible act of amplification' (see page 156). The truth of the matter is that we gain information about the microscopic world only when we can amplify elementary quantum events like the absorption of photons, and turn them into perceptible macroscopic signals involving the deflection of a pointer on a scale, etc. Is this process of bridging between the microworld and the macroworld a logical place for the collapse of the wavefunction? If so, Schrödinger's cat might then be spared at least the discomfort of being both dead and alive, because the act of amplification associated with the registering of a radioactive emission by the Geiger counter settles the issue before a superposition of macroscopic states can be generated.

However, as we have repeatedly stressed in this book, the Copenhagen interpretation of quantum theory leaves unanswered the question of just where the collapse of the wavefunction takes place. John Bell wrote of the 'shifty split' between measured object and perceiving subject:[‡] 'What exactly qualifies some physical systems to play the role of

---

[†] Prigogine, Ilya (1980). *From being to becoming*. W. H. Freeman, San Francisco, CA.
[‡] Bell, J. S. (1990). *Physics World*. 3, 33.

"measurer"? Was the wavefunction of the world waiting to jump for thousands of years until a single-celled living creature appeared? Or did it have to wait a little longer, for some better qualified system . . . with a PhD?' Bell argues that one way to avoid the 'shifty split' is to introduce some extra element in the theory which ensures that the wavefunction is effectively collapsed at a very early stage in the process of amplification.

Is such an extension possible? The Italian physicists G.C. Ghiradi, A. Rimini and T. Weber (GRW) formulated just such a theory in 1986. To the usual non-relativistic, time-symmetric equations of motion, they added a non-linear term which subjects the wavefunction to random, spontaneous localizations in configuration space. Their ambition was primarily to bridge the gap between the dynamics of microscopic and macroscopic systems in a unified theory. To achieve this, they introduced two new constants whose orders of magnitude were chosen so that (i) the theory does not contradict the usual quantum theory predictions for microscopic systems, (ii) the dynamical behaviour of a macroscopic system can be derived from its microscopic constituents and is consistent with classical dynamics, and (iii) the wavefunction is collapsed by the act of amplification, leading to well defined individual macroscopic states of pointers and cats, etc.

One of these new constants represents the frequency of spontaneous localizations of the wavefunction. For a microscopic system (an individual quantum particle or a small collection of such particles) GRW chose for the localization frequency a value of $10^{-16} \, \text{s}^{-1}$. This implies that the wavefunction is localized about once every billion years. In practical terms, the wavefunction of a microscopic system *never* localizes: it continues to evolve in time according to the time-symmetric equations of motion derived from the time-dependent Schrödinger equation. There is therefore no practical difference between the GRW theory and orthodox quantum theory for microscopic systems. However, for macroscopic systems GRW suggest a localization frequency of $10^7 \, \text{s}^{-1}$; i.e. the wavefunction is localized within about 100 nanoseconds. The difference between these two frequencies is simply related to the *number* of particles involved.

Because a measuring device is a large object like a photomultiplier (or a cat), the wavefunction is collapsed in the very early stages of the measurement process. Bell wrote that in the GRW extension of quantum theory, '[Schrödinger's] cat is not both dead and alive for more than a split second.'

We should note that the GRW theory serves only to sharpen the collapse of the wavefunction and make it a necessary part of the process of amplification. It does not solve the need to invoke the 'spooky action at a distance' implied by the results of the Aspect experiments described in Chapter 4. The GRW theory would predict that in those experiments, the detection and amplification of either photon automatically collapses the whole (spatially quite delocalized) wavefunction. The properties of the other, not yet detected, photon change from being possibilities into actualities at the moment this collapse takes place. Bell himself demonstrated that this action at a distance need not imply that 'messages' must be sent between the photons and that, therefore, there is nothing in the GRW theory to contradict the demands of special relativity. In fact, although GRW originally formulated their theory as an extension of non-relativistic quantum mechanics, they have now generalized it to include the effects of special relativity and can apply it to systems containing identical particles.

Macroscopic quantum objects

Of course, in the 55 years since Schrödinger first introduced the world to his cat, no-one has ever reported seeing a cat in a linear superposition state (at least, not in a reputable scientific journal). The GRW theory suggests that such a thing is impossible because the wavefunction collapses much earlier in the measurement process. However, the theory could run into difficulties if linear superpositions of *some* kinds of macroscopic quantum states could be generated in the laboratory.

Every undergraduate scientist knows that particles of like charge repel one another. However, when cooled to very low temperatures, two electrons moving through a lattice of metal ions can experience a small mutual *attraction* which is greatest when they possess opposite spin orientations. This attraction is indirect: one electron interacts with the lattice of metal ions and deforms it slightly. The second electron senses this deformation and can reduce its energy in response. The result is an attraction between the two electrons mediated by the lattice deformation.

Electrons are fermions and obey the Pauli exclusion principle (see Section 2.4), but when considered as though they are a single entity, two spin-paired electrons have no net spin and so collectively form a boson. Like other bosons (such as photons), these pairs of electrons can 'condense' into a single quantum state. When a large number of pairs so condense, the result is a macroscopic quantum state extending over large distances (i.e. several centimetres). In this condensed state, which lies lower in energy than the normal conduction band of the metal, the electrons experience no resistance. This is the superconducting state of the metal.

The attraction between the electrons is very weak and is easily overcome by thermal motion (hence the need for very low temperatures). The distance between each electron in a pair is consequently quite large, and so many such pairs overlap within the metal lattice. The wavefunctions of the pairs likewise overlap and their peaks and troughs line up just like light waves in a laser beam. The result can be a macroscopic number (10^{20}) of electrons moving through a metal lattice with their individual wavefunctions locked in phase.

The attentions of theoretical and experimental physicists have focused on the properties of superconducting rings. Imagine that an external magnetic field is applied to a metal ring, which is then cooled to its superconducting temperature. The current which flows in the surface of the ring forces the magnetic field to flow outside the body of the material. The total field is just the sum of the applied field and the field induced by the current flowing in the surface of the ring. If the applied field is removed, the current continues to circulate (because the electrons feel

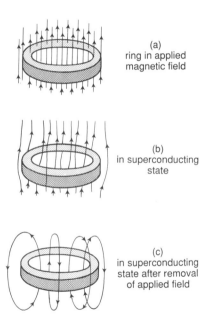

(a)
ring in applied
magnetic field

(b)
in superconducting
state

(c)
in superconducting
state after removal
of applied field

Fig. 5.6 (a) A superconducting ring is placed in a magnetic field and cooled to its superconducting temperature. (b) In the superconducting state, pairs of electrons pass through the metal with no resistance and the magnetic field is forced to pass around the outside of the body of the ring. (c) When the applied magnetic field is removed, the superconducting electrons generate a 'trapped' magnetic flux which is quantized. From Richard P. Feynman, *The Feynman Lectures on physics* Vol. III © 1965 California Institute of Technology. Reprinted with permission of Addison Wesley Publishing Company, Inc.

no resistance) and an amount of magnetic flux is 'trapped', as shown in Fig. 5.6. According to the quantum theory of superconductivity, this trapped flux is quantized: only integer multiples of the so-called super-conducting flux quantum (given by $h/2e$, where e is the electron charge) are allowed. These different flux states therefore represent the quantum states of an object of macroscopic dimensions — such superconducting rings are usually about a centimetre or so in diameter. The existence of these states has been confirmed by experiment.

In a superconducting ring of uniform thickness, the quantized magnetic flux states do not interact. The quantum state of the ring can only be changed by warming it up, changing the applied external field and then cooling it down to its superconducting temperature again. However, the mixing of the flux states becomes possible if the ring has a so-called weak link. This is essentially a small region of the ring where the thickness of the material is reduced to about a few hundred ångströms and across which the magnetic flux quanta can 'leak'. These objects have a wide variety of macroscopic quantum mechanical properties which have been explored experimentally despite the difficulties associated with the need to make sensitive measurements at very low temperatures. Most impor-tantly, the instruments used to make these sensitive measurements may be smaller than the macroscopic quantum object being studied. It is therefore possible to make non-invasive measurements, in which the object remains in the same eigenstate throughout.

Perhaps of greatest relevance to the present discussion is the use of these objects as superconducting quantum interference devices (SQUIDs). Many different types of quantum effects have been demon-strated using these devices (including quantum 'tunnelling'), but one of the most interesting experiments has yet to be performed successfully. This experiment involves the generation of a linear superposition of macroscopically different SQUID states. For example, a superposition of two states in which macroscopic numbers of electrons flow around the ring *in opposite directions*. Such a superposition, which is routine in the quantum world, would contradict our basic understanding of the macroscopic world in which large objects are seen to be in one state or the other but not both simultaneously. Although a SQUID is not Schrödinger's cat, it is at least one of its close cousins. If they can ever be performed, these experiments could have a profound impact on our understanding of the quantum measurement process.

Quantum gravity

The mathematical fusion of quantum theory and special relativity into quantum field theory was fraught with difficulties. The mathematics tended to produce irritating infinities which were eventually removed through the process of renormalization – a process still regarded by some physicists as rather unsatisfactory. However, these difficulties pale into insignificance compared with those encountered when attempts are made to fuse quantum field theory with general relativity. If this merging of the two most successful of physical theories could ever be achieved, the result would be a theory of quantum gravity. To date, we are not even close to such a theory. Indeed, physicists and mathematicians do not even know what a theory of quantum gravity should look like.

In Einstein's general theory of relativity, the action at a distance implied by the classical (Newtonian) force of gravity is replaced by a curved space–time. The amount of curvature in a particular region of space–time is related to the density of mass and energy present (since $E = mc^2$). Calculating this density is no problem in classical physics, but in quantum theory the momentum of an object is replaced by a differential operator ($\hat{p}_x = -i\hbar\partial/\partial x$), leading to an immediate problem of interpretation.

There are other, much more profound problems, however. In quantum field theory, quantum fluctuations can give rise to the creation of 'virtual' particles out of nothing (the vacuum), provided that the particles mutually annihilate before violating the uncertainty principle. Now when Einstein first developed his general theory of relativity, he introduced a 'fudge' factor which he called the cosmological constant (he had his reasons). This constant he later withdrew from the theory but, in fact, it turns out to be related to the energy density of the vacuum that results from quantum fluctuations. Quantum field theory – in the form known to physicists as the standard model – makes predictions for some of the contributions to the cosmological constant, and estimates can be made of the others. The standard model says that this constant should be sizeable: observations on distant parts of the universe say the constant is effectively zero.

In fact, if the cosmological constant had the value suggested by the standard model, the curvature of space–time would be visible to *us*, and the world would look very strange indeed. Either some impressive cancellation of terms is responsible, or the theory is flawed. One possibility, originally put forward by the mathematician Stephen Hawking and developed by Sidney Coleman, is that the quantum fluctuations create a myriad of 'baby universes', connected to our own universe by quantum wormholes with widths given by the so-called Planck length, 10^{-33} cm. The wormholes would look like tiny black holes, flickering in to and out of existence within 10^{-43} s. Coleman has suggested that such wormholes could cancel the contributions to the cosmological constant made by the particle fields. This theory has some way to go but, rather interestingly, it has been claimed that experimental tests might be possible using a SQUID.

The mathematician Roger Penrose believes that if these difficulties can be overcome, the resulting theory of quantum gravity will provide a solution to the problem of the collapse of the wavefunction. Some extra ingredients will have to be added, however, since both quantum theory and general relativity are time symmetric. Nevertheless, Penrose has argued that a linear superposition of quantum states will begin to break down and eventually collapse into a specific eigenstate when a region of significant space–time curvature is entered. Unlike the GRW theory, in which the number of particles is the key to the collapse, in Penrose's theory it is the density of mass–energy which is important.

Gravitational effects are unlikely to be very significant at the microscopic level of individual atoms and molecules, and so the wavefunction is expected to evolve in the usual time-symmetric fashion according to the dynamical equations of quantum theory. Penrose suggests that it is at the level of one graviton where the curvature of space–time becomes sufficient to ensure the time-asymmetric collapse of the wavefunction. The graviton is the (as yet unseen) quantum particle of the gravitational field, much like the photon is the quantum particle of the electromagnetic field. It is associated with a scale of mass known as the *Planck mass*, about 10^{-5} g. This is rather a large mass requirement to trigger the collapse. Penrose has responded by further suggesting that it is the *difference* between gravitational fields in the space–times of different measurement possibilities which is important. This difference can quickly exceed one graviton, forcing the wavefunction to collapse into one of the eigenstates.

Penrose accepts that this is merely the germ of an idea which needs to be pursued much further. In his recent book *The emperor's new mind*, first published in 1989, he writes:[†]

It is my opinion that our present picture of physical reality, particularly in relation to the nature of *time*, is due for a grand shake-up — even greater, perhaps, than that which has already been provided by present-day relativity and quantum mechanics.

Theories of quantum gravity and quantum cosmology are in their infancy and many speculative proposals have been made. It is certainly true that although we have come an awfully long way, there are still huge gaps in our understanding of time, the universe and its constitent bits and pieces. Consequently, a technical solution to the quantum measurement problem — one which emerges from some new theory in an entirely objective manner — may eventually be found, and we have examined some possible candidates in this section. Other approaches to the problem

---

[†] Penrose, Roger, (1990). *The emperor's new mind*. Vintage, London.

have been taken, however, and we will now turn our attention to some of these.

5.3 THE CONSCIOUS OBSERVER

Be warned, in the last three sections of this chapter we are going to leave what might appear to be the straight and narrow paths of science and wander in the realms of metaphysical speculation. Of course, what we have been discussing so far in this chapter has not been without its metaphysical elements, but at least the attempts described above to make quantum theory more objective are expressed in the language most scientists feel at home with. Before we plunge in at the deep end, perhaps we should review briefly the steps that have led us here.

The orthodox Copenhagen interpretation of quantum theory is silent on the question of the collapse of the wavefunction. The field is therefore wide open. If we choose to reject the strict Copenhagen interpretation we are, given our present level of understanding, free to choose exactly how we wish to fill the vacuum. Any suggestion, no matter how strange, is acceptable provided that it does not produce a theory inconsistent with the predictions of quantum theory known to have been so far upheld by experiment. Our choice is a matter of personal taste.

Now we can try to be objective about how we change the theory to make the collapse explicit, and the GRW theory and quantum gravity are good examples of that approach. But we should remember that there is no a priori reason why we should distinguish between the observed quantum object and the measuring apparatus based on size or the curvature of space–time or any other inherent physical property, other than the fact that we seem to possess a theory of the microscopic world that sits very uncomfortably in our macroscopic world of experience. However, macroscopic measuring devices are undisputably made of microscopic quantum particles, and should therefore obey the rules of quantum theory unless we add something to the theory specifically to change those rules. If the consequences were not so bizarre we would, perhaps, have no real difficulty in accepting that quantum theory should be no less applicable to large objects than to atoms and molecules. In fact, this was something that John von Neumann was perfectly willing to accept.

Von Neumann's theory of measurement

John von Neumann's *Mathematical foundations of quantum mechanics* was an extraordinarily influential work. It is important to note that the language we have used in this book to describe and discuss the measurement process in terms of a collapse or projection of the wave-function essentially originates with this classic book. It was von Neumann who so clearly distinguished (in the mathematical sense) between the continuous time-symmetric quantum mechanical equations of motion and the discontinuous, time-asymmetric measurement process. Although much of his contribution to the development of the theory was made within the boundaries of the Copenhagen view, he stepped beyond those boundaries in his interpretation of quantum measurement.

Von Neumann saw that there was no way he could obtain an irreversible collapse of the wavefunction from the equations of quantum theory. Yet he demonstrated that if a quantum system is present in some eigenstate of a measuring device, the product of this eigenstate and the state vector of the measuring device should evolve in time in a manner quite consistent with both the quantum mechanical equations of motion and the expected measurement probabilities. In other words, there is no mathematical reason to suppose that quantum theory does not account for the behaviour of macroscopic measuring devices. This is where von Neumann goes beyond the Copenhagen interpretation.

So how does the collapse of the wavefunction arise? Von Neumann's book was published in German in Berlin in 1932, three years before the publication of the paper in which Schrödinger introduced his cat. The problem is this: unless it is supposed that the collapse occurs somewhere in the measurement process, we appear to be stuck with an infinite regress and with animate objects suspended in superposition states of life and death. Von Neumann's answer was as simple as it is alarming: the wavefunction collapses when it interacts with a *conscious observer*.

It is difficult to fault the logic behind this conclusion. Quantum particles are known to obey the laws of quantum theory: they are described routinely in terms of superpositions of the measurement eigenstates of devices designed to detect them. Those devices are themselves composed of quantum particles and should, in principle, behave similarly. This leads us to the presumption that linear superpositions of macroscopically different states of measuring devices (different pointer positions, for example) are possible. But the observer never actually sees such superpositions.

Von Neumann argued that photons scattered from the pointer and its scale enter the eye of the observer and interact with his retina. This is still a quantum process. The signal which passes (or does not pass) down the observer's optic nerve is in principle still represented in terms of a linear superposition. Only when the signal enters the brain and thence the conscious mind of the observer does the wavefunction encounter a 'system' which we can suppose is not subject to the time-symmetrical laws of quantum theory, and the wavefunction collapses. We still have a basic dualism in nature, but now it is a dualism of *matter* and conscious *mind*.

Wigner's friend

But *whose* mind? In the early 1960s, the physicist Eugene Wigner addressed this problem using an argument based on a measurement made through the agency of a second observer. This argument has become known as the paradox of Wigner's friend.

Wigner reasoned as follows. Suppose a measuring device is constructed which produces a flash of light every time a quantum particle is detected to be in a particular eigenstate, which we will denote as $|\psi_+\rangle$. The corresponding state of the measuring device (the one giving a flash of light) is denoted $|\phi_+\rangle$. The particle can be detected in one other eigenstate, denoted $|\psi_-\rangle$, for which the corresponding state of the measuring device (no flash of light) is $|\phi_-\rangle$. Initially, the quantum particle is present in the superposition state $|\Psi\rangle = c_+ |\psi_+\rangle + c_- |\psi_-\rangle$. The combination (particle in state $|\psi_+\rangle$, light flashes) is given by the product $|\psi_+\rangle |\phi_+\rangle$. Similarly the combination (particle in state $|\psi_-\rangle$, no flash) is given by the product $|\psi_-\rangle |\phi_-\rangle$. If we now treat the combined system—particle plus measuring device—as a single quantum system, then we must express the state vector of this combined system as a superposition of the two possibilities: $|\Phi\rangle = c_+ |\psi_+\rangle |\phi_+\rangle + c_- |\psi_-\rangle |\phi_-\rangle$ (see the discussion of entangled states and Schrödinger's cat in Section 3.4).

Wigner can discover the outcome of the next quantum measurement by waiting to see if the light flashes. However, he chooses not to do so. Instead, he steps out of the laboratory and asks his friend to observe the result. A few moments later, Wigner returns and asks his friend if he saw the light flash.

How should Wigner analyse the situation before his friend speaks? If he now considers his friend to be part of a larger measuring 'device', with states $|\phi'_+\rangle$ and $|\phi'_-\rangle$, then the total system of particle plus measuring device plus friend is represented by the superposition state $|\Phi'\rangle = c_+ |\psi_+\rangle |\phi'_+\rangle + c_- |\psi_-\rangle |\phi'_-\rangle$. Wigner can therefore anticipate that there will be a probability $|c_+|^2$ that his friend will answer 'Yes' and a probability $|c_-|^2$ that he will answer 'No'. If his friend answers 'Yes', then as far as Wigner himself is concerned the wavefunction $|\Phi'\rangle$ collapses at that moment and the probability that the alternative result was obtained is reduced to zero. Wigner thus infers that the particle was detected in the eigenstate $|\psi_+\rangle$ and that the light flashed.

But now Wigner probes his friend a little further. He asks 'What did you feel about the flash before I asked you?', to which his friend replies: 'I told you already, I did [did not] see a flash.' Wigner concludes (not unreasonably) that his friend must have already made up his mind about the measurement before he was asked about it. Wigner wrote that the state vector $|\Phi'\rangle$ '. . . appears absurd because it implies that my friend was in a state of suspended animation before he answered my question.'[†] And yet we know that if we replace Wigner's friend with a simple physical system such as a single atom, capable of absorbing light from the flash, then the mathematically correct description is in terms of the superposition $|\Phi'\rangle$, and not either of the collapsed states $|\psi_+\rangle\,|\phi'_+\rangle$ or $|\psi_-\rangle\,|\phi'_-\rangle$. 'It follows that the being with a consciousness must have a different role in quantum mechanics than the inanimate measuring device: the atom considered above.'[†] Of course, there is nothing in principle to prevent Wigner from assuming that his friend was indeed in a state of suspended animation before answering the question. 'However, to deny the existence of the consciousness of a friend to this extent is surely an unnatural attitude.'[†] That way also lies solipsism — the view that all the information delivered to your conscious mind by your senses is a figment of your imagination, i.e. nothing exists but your consciousness.

Wigner was therefore led to argue that the wavefunction collapses when it interacts with the *first* conscious mind it encounters. Are cats conscious beings? If they are, then Schrödinger's cat might again be spared the discomfort of being both alive and dead: its fate is already decided (by its own consciousness) before a human observer lifts the lid of the box.

Conscious observers would therefore appear to violate the physical laws which govern the behaviour of inanimate objects. Wigner calls on a second argument in support of this view. Nowhere in the physical world is it possible physically to act on an object without some kind of reaction. Should consciousness be any different? Although small, the action of a conscious mind in collapsing the wavefunction produces an immediate reaction — knowledge of the state of a system is irreversibly (and indelibly) generated in the mind of the observer. This reaction may lead to other physical effects, such as the writing of the result in a laboratory notebook or the publication of a research paper. In this hypothesis, the influence of matter over mind is balanced by an influence of mind over matter.

[†] Wigner, Eugene in Good, I. J. (ed.) (1961). *The scientist speculates: an anthology of partly-baked ideas*. Heinemann, London.

The ghost in the machine

This cannot be the end of the story, however. Once again, we see that a proposed solution to the quantum measurement problem is actually no solution at all—it merely shifts the focus from one thorny problem to another. In fact, the approach adopted by von Neumann and Wigner forces us to confront one of philosophy's oldest problems: what is consciousness? Just how does the consciousness (mind) of an observer relate to the corporeal structure (body) with which it appears to be associated?

Although our bodies are outwardly different in appearance, it is our consciousness that allows us to perceive ourselves as individuals and to relate that sense of self to the world outside. Consciousness defines who we *are*. It is the storehouse for our memories, thoughts, feelings and emotions and governs our personality and behaviour.

The seventeenth century philosopher René Descartes chose consciousness as the starting point for what he hoped would become a whole new philosophical tradition. In his *Discourse on method*, published in 1637, he spelled out the criteria he had set for himself in establishing a rigorous approach based on the apparently incontrovertible logic of geometry and mathematics. He would accept nothing that could be doubted: '. . . as I wanted to concentrate solely on the search for truth, I thought I ought to . . . reject as being absolutely false everything in which I could suppose the slightest reason for doubt . . .'[†] In this way, he could build his new philosophical tradition with confidence in the absolute truth of its statements. This meant rejecting information about the world received through his senses, since our senses are easily deceived and therefore not to be trusted.

Descartes argued that as he thinks independently of his senses, the very fact that he thinks is something about which he can be certain. He further concluded that there is an essential contradiction in holding to the belief that something that thinks does not also exist, and so his existence was also something about which he could be certain. *Cogito ergo sum*, he concluded—I think therefore I am.

While Descartes could be confident in the truth of his existence as a conscious entity, he could not be confident about the appearances of things revealed to his mind by his senses. He therefore went on to reason that the thinking 'substance' (consciousness or mind) is quite distinct from the unthinking 'machinery' of the body. The machine is just another form of extended matter (it has extension in three-dimensional space) and may—or may not—exist, whereas the mind has no extension

---

[†] Descartes, René (1968). *Discourse on method and the meditations*. Penguin, London.

and must exist. Descartes had to face up to the difficult problem of deciding how something with no extension could influence and direct the machinery — how a thought could be translated into movement of the body. His solution was to identify the pineal gland, a small pear-shaped organ in the brain, as the 'seat' of consciousness through which the mind gently nudges the body into action.

This mind–body dualism (Cartesian dualism) in Descartes's philosophy is entirely consistent with the medieval Christian belief in the soul or spirit, which was prevalent at the time he published his work. The body is thus merely a shell, or host, or mechanical device used for giving outward expression and extension to the unextended thinking substance. My mind defines who I *am* whereas my body is just something I *use* (perhaps temporarily). Descartes believed that although mind and body are joined together, connected through the pineal gland, they are quite capable of separate, independent existence. In his seminal book *The concept of mind*, Gilbert Ryle wrote disparagingly of this dualist conception of mind and body, referring to it as the 'ghost in the machine'.

Now Descartes's reasoning can, and has been, heavily criticized. He had wanted to establish a new philosophical tradition by adhering to some fairly rigorous criteria regarding what he could and could not accept to be beyond doubt. And yet his most famous statement — 'I think therefore I am' — was arrived at by a process which seems to involve assumptions that, by his own criteria, appear to be unjustified. The statement is also a linguistic nightmare and, as the logical positivists later demonstrated to their obvious satisfaction, consequently quite without meaning.

Has our understanding of consciousness improved since the seventeenth century? Certainly, we now know a great deal more about the functioning of the brain. We know how various parts of the body and various activities (such as speech) are controlled by different parts of the brain. We know something about the brain's chemistry and physiology; for example, we now associate many 'mental' disorders with hormone imbalances. We know quite a lot more about the *machinery*, but twentieth century science appears to have taken us no closer to *mind*. What has changed is that modern scientists tend to regard the mind not as Descartes's separate, unextended thinking substance capable of independent existence, but as a natural product of the complex machinery of the brain. However, mind continues to be more of a subject for philosophical, rather than scientific, inquiry.

Brain states and quantum memory

If the brain is just a complicated machine, then presumably it acts just like another measuring device, as von Neumann reasoned. In fact, the dark-adapted eye is a very good example of a detection device capable of operating at the quantum level. It can respond to the absorption of a single photon by the retina. We do not 'see' single photons because the brain has a mechanism for filtering out such weak signals as peripheral 'noise' (but we can see as few as 10 photons if they arrive together). The wavefunction of the photons and what we might consider as the 'wavefunction of the brain' presumably combine in a superposition state which is then somehow collapsed by the mind. It follows that, prior to the collapse, the brain of a conscious observer exists in a superposition of states.

There is an alternative possibility. What if the wavefunction does not collapse at all and, instead, the 'stream of consciousness' of the observer is split by the measurement process? In his book *Mind, brain and the quantum: the compound 'I'*, Michael Lockwood puts forward the proposal that the consciousness of the observer enters a superposition state. Each of the different measurement possibilities are therefore realized, registered in different versions of the observer's conscious mind. Presumably, each version will be statistically weighted according to the modulus-squares of the projection amplitudes in the usual way. But the observer is aware of, and remembers, only one result.

The observer has, in principle, a kind of quantum memory of the measurement process in which different possibilities are recalled in different parallel states of consciousness. Over time, we might expect these parallel selves to develop into distinctly different individuals as a multitude of quantum events washes over the observer's senses. Within one brain may be not one, but many ghosts.

Free will and determinism

You may have been tempted from time to time in your reading of this book to cast your mind back to the good old days of Newtonian physics where everything seemed to be set on much firmer ground. Classical physics was based on the idea of a grand scheme: a mechanical clockwork universe where every effect could be traced back to a cause. Set the clockwork universe in motion under some precisely known initial conditions and it should be possible to predict its future development in unlimited detail.

However, apart from the reservations we now have about our ability to know the initial conditions with sufficient precision, there are two fairly profound philosophical problems associated with the idea of a completely deterministic universe. The first is that if every effect must have a cause, then there must have been a *first* cause that brought the universe into existence. The second is that, if every effect is determined by the behaviour of material entities conforming to physical laws, what happens to the notion of *free will*? We will defer discussion of the first problem until Section 5.5, and turn our attention here to the second problem.

The Newtonian vision of the world is essentially reductionist: the behaviour of a complicated object is understood in terms of the properties and behaviour of its elementary constituent parts. If we apply this vision to the brain and consciousness, we are ultimately led to the modern view that both should be understood in terms of the complex (but deterministic) physical and chemical processes occurring in the machinery. Taken to its extreme, this view identifies the mind as the software which programmes the hardware of the brain. The proponents of so-called strong AI (artificial intelligence) believe that it should one day be possible to develop and programme a computer to *think*.

One consequence of this completely deterministic picture is that our individual personalities, behaviour, thoughts, actions, emotions, etc., are effects which we should in principle be able to trace back to a set of one or more material causes. For example, my choice of words in this sentence is not a matter for my individual freedom of will, it is a necessary consequence of the many physical and chemical processes occurring in my brain. That I should decide to boldly split an infinitive in this sentence was, in principle, dictated by my genetic makeup and physical environment, integrated up to the moment that my 'state of mind' led me to 'make' my decision.

We should differentiate here between actions that are essentially instinctive (and which are therefore *re*actions) and actions based on an apparent freedom of choice. I would accept that my reaction to pain is entirely predictable, whereas my senses of value, justice, truth and beauty seem to be matters for me to determine as an individual. Ask an individual exhibiting some pattern of conditioned behaviour, and he will tell you (somewhat indignantly) that, at least as far as accepted standards of behaviour and the law are concerned, he has his own mind and can exercise his own free will. Is he suffering a delusion?

Before the advent of quantum theory, the answer given by the majority of philosophers would have been 'Yes'. As we have seen, Einstein himself was a realist and a determinist and consequently rejected the idea of free will. In choosing at some apparently unpredictable moment to light his pipe, Einstein saw this not as an expression of his freedom to will a certain action to take place, but as an effect which has some physical cause. One possible explanation is that the chemical balance of his brain is upset by a low concentration of nicotine, a chemical on which it had come to depend. A complex series of chemical changes takes place which is translated by his mind as a desire to smoke his pipe. These chemical changes therefore cause his mind to will the act of lighting his pipe, and that act of will is translated by the brain into bodily movements designed to achieve the end result. If this is the correct view, we are left with nothing but physics and chemistry.

In fact, is it not true that we tend to analyse the behaviour patterns of everyone (with the usual exception of ourselves) in terms of their personalities and the circumstances that lead to their acts. Our attitude towards an individual may be sometimes irreversibly shaped by a 'first impression', in which we analyse the physiognomy, speech, body language and attitudes of a person and come to some conclusion as to what 'kind' of person we are dealing with. How often do we say: 'Of course, that's just what you would expect him to do in those circumstances'? If we analyse our own past decisions carefully, would we not expect to find that the outcomes of those decisions were entirely predictable, based on what we know about ourselves and our circumstances at the time? Is anyone truly *un*predictable?

Classical physics paints a picture of the universe in which we are nothing but fairly irrelevant cogs in the grand machinery of the cosmos. However, quantum physics paints a rather different picture and may allow us to restore some semblance of self-esteem. Out go causality and determinism, to be replaced by the indeterminism embodied in the uncertainty relations. Now the future development of a system becomes impossible to predict except in terms of probabilities. Furthermore, if we accept von Neumann's and Wigner's arguments about the role of consciousness in quantum physics, then our conscious selves become the most important things in the universe. Quite simply, without conscious observers, there would be no physical reality. Instead of tiny cogs forced to grind on endlessly in a reality not of our design and whose purpose we cannot fathom, we become the *creators* of the universe. *We* are the masters.

However, we should not get too carried away. Despite this changed role, it does not necessarily follow that we have much freedom of choice in quantum physics. When the wavefunction collapses (or when Lockwood's conscious self splits), it does so unpredictably in a manner which would seem to be beyond our control. Although our minds may be essential to the realization of a particular reality, we cannot know or decide in advance what the result of a quantum measurement will be. We cannot choose what kind of reality we would like to perceive beyond choosing the measurement eigenstates. In this interpretation of quantum measurement, our only influence over matter is to make it real. Unless we are prepared to accept the possibility of a variety of paranormal phenomena, it would seem that we cannot bend matter to our will.

Of course, the notion that a conscious mind is necessary to sustain reality is not new to philosophers, although it is perhaps a novel experience to find it advocated as a key explanation in one of the most important and successful of twentieth century scientific theories.

5.4 THE 'MANY-WORLDS' INTERPRETATION

The concept of the collapse of the wavefunction was introduced by von Neumann in the early 1930s and has become an integral part of the orthodox interpretation of quantum theory. What evidence do we have that this collapse really takes place? Well . . . none, actually. The collapse is necessary to explain how a quantum system initially present in a linear superposition state before the process of measurement is converted into a quantum system present in one, and only one, of the measurement eigenstates after the process has occurred. It was introduced into the theory because it is our experience that pointers point in only one direction at a time.

The Copenhagen solution to the measurement problem is to say that there is no solution. Pointers point because they are part of a macroscopic measuring device which conforms to the laws of classical physics. The collapse is therefore the only way in which the 'real' world of classical objects can be related to the 'unreal' world of quantum particles. It is simply a useful invention, an algorithm, that allows us to predict the outcomes of measurements. As we pointed out in the previous two sections, if we wish to make the collapse a real physical change occurring in a real physical property of a quantum system, then we must *add* something to the theory, if only the suggestion that consciousness is somehow involved.

The simplest solution to the problem of quantum measurement is to say that there is no problem. Over the last 60 years, quantum theory has proved its worth time and time again in the laboratory: why change it or add extra bits to it? Although it is overtly a theory of the microscopic world, we know that macroscopic objects are composed of atoms and molecules, so why not accept that quantum theory applies equally well to pointers, cats and human observers? Finally, if we have no evidence for the collapse of the wavefunction, why introduce it?

In Lockwood's interpretation described in the last section, the observer was assumed to split into a number of different, non-interacting conscious selves. Each individual self records and remembers a different result, and *all* results are realized. In fact, Lockwood's approach is closely related to an older interpretation proposed over 30 years ago by Hugh Everett III in his Princeton University Ph.D. thesis. In this interpretation the act of measurement splits the *entire universe* into a number of branches, with a different result being recorded in each. This is the so-called 'many-worlds' interpretation of quantum theory.

Relative states

Everett discussed his original idea, which he called the 'relative state' formulation of quantum mechanics, with John Wheeler while at Princeton. Wheeler encouraged Everett to submit his work as a Ph.D. thesis, which he duly did in March 1957. A shortened version of this thesis was published in July 1957 in the journal *Reviews of Modern Physics*, and was 'assessed' by Wheeler in a short paper published in the same issue. Everett set out his interpretation in a much more detailed article which was eventually published in 1973, together with copies of some other relevant papers, in the book *The many-worlds interpretation of quantum mechanics*, edited by Bryce S. DeWitt and Neill Graham. Everett's original work was largely ignored by the physics community until DeWitt and Graham began to look at it more closely and to popularize it some 10 years later.

Everett insisted that the pure Schrödinger wave mechanics is all that is needed to make a complete theory. Thus, the wavefunction obeys the deterministic, time-symmetric equations of motion at all times in all circumstances. Initially, no interpretation is given for the wavefunction; rather, the meaning of the wavefunction emerges from the formalism itself. Without the collapse of the wavefunction, the measurement process occupies no special place in the theory. Instead, the results of the interaction between a quantum system and an external observer are obtained from the properties of the larger composite system formed from them.

In complete contrast to the special role given to the observer in von Neumann's and Wigner's theory of measurement, in Everett's interpretation the observer is nothing more than an elaborate measuring device. In terms of the effect on the physics of a quantum system, a conscious observer is no different from an inanimate, automatic recording device which is capable of storing an experimental result in its memory.

The 'relative state' formulation is based on the properties of quantum systems which are composites of smaller sub-systems. Each sub-system can be described in terms of some state vector which, in turn, can be written as a linear superposition of some arbitrary set of basis states. Thus, each sub-system is described by a set of basis state vectors in an associated Hilbert space. The Hilbert space of the composite system is the *tensor product* of the Hilbert spaces of the sub-systems. If we consider the simple case of two sub-systems, the overall state vector of the composite is a grand linear superposition of terms in which each element in the superposition of one sub-system multiplies every element in the superposition of the other. The end result is equivalent to that given in our discussion of entangled states in Section 3.4.

We can see more clearly what this means by looking at a specific example. Let us consider once again the interaction between a simple quantum system and a measuring device for which the system possesses just two eigenstates. The measuring device may, or may not, involve observation by a human observer. Following from our previous discussions, we can write the state vector of the composite system (quantum system plus measuring device) as $|\Phi\rangle = c_+ |\psi_+\rangle |\phi_+\rangle + c_- |\psi_-\rangle |\phi_-\rangle$. As before $|\psi_+\rangle$ and $|\psi_-\rangle$ are the measurement eigenstates of the quantum system and $|\phi_+\rangle$ and $|\phi_-\rangle$ are the corresponding states of the measuring device (different pointer positions, for example) after the interaction has taken place. Everett's argument is that we can no longer speak of the state of either the quantum system or the measuring device independently of the other. However, we can define the states of the measuring device *relative* to those of the quantum system as follows:

$$|\Phi_{REL}^+\rangle = c_+ |\phi_+\rangle$$
$$|\Phi_{REL}^-\rangle = c_- |\phi_-\rangle \tag{5.3}$$

whence

$$|\Phi\rangle = |\psi_+\rangle |\Phi_{REL}^+\rangle + |\psi_-\rangle |\Phi_{REL}^-\rangle. \tag{5.4}$$

The relative nature of these states is made more explicit by writing the expansion coefficients c_+ and c_- as the projection amplitudes:

$$c_+ = \langle \psi_+, \phi_+ | \Phi \rangle, |\Phi_{REL}^+\rangle = \langle \psi_+, \phi_+ | \Phi \rangle |\phi_+\rangle$$
$$c_- = \langle \psi_-, \phi_- | \Phi \rangle, |\Phi_{REL}^-\rangle = \langle \psi_-, \phi_- | \Phi \rangle |\phi_-\rangle \tag{5.5}$$

where $\langle \psi_+, \phi_+ | = \langle \psi_+ | \langle \phi_+ |$ and $\langle \psi_-, \phi_- | = \langle \psi_- | \langle \phi_- |$.

Everett went on to show that his relative states formulation of quantum mechanics is entirely consistent with the way quantum theory is used in its orthodox interpretation to derive probabilities. Instead of talking about projection amplitudes and probabilities, it is necessary to talk about conditional probabilities: the probability that a particular result will be obtained in a measurement given certain conditions. The name is different, but the procedure is the same.

All this is reasonably straightforward and non-controversial. However, the logical extension of Everett's formulation of quantum theory leads inevitably to the conclusion that, once entangled, the relative states can *never* be disentangled.

The branching universe

In Everett's formulation of quantum theory, there is no doubt as to the reality of the quantum system. Indeed, his theory is quite deterministic in the way that Schrödinger had originally hoped that his wave mechanics was deterministic. Given a certain set of initial conditions, the wavefunction of the quantum system develops according to the quantum laws of (essentially wave) motion. The wavefunction describes the real properties of a real system and its interaction with a real measuring device: all the speculation about determinism, causality, quantum jumps and the collapse of the wavefunction is unnecessary. However, the restoration of reality in Everett's formalism comes with a fairly large trade-off. If there is no collapse, each term in the superposition of the total state vector $|\Phi\rangle$ is real — *all* experimental results are realized.

Each term in the superposition corresponds to a state of the composite system and is an eigenstate of the observation. Each describes the correlation of the states of the quantum system and measuring device (or observer) in the sense that $|\psi_+\rangle$ is correlated with $|\phi_+\rangle$ and $|\psi_-\rangle$ with $|\phi_-\rangle$. Everett argued that this correlation indicates that the observer perceives only one result, corresponding to a specific eigenstate of the observation. In his July 1957 paper, he wrote:[†]

Thus with each succeeding observation (or interaction), the observer state 'branches' into a number of different states. Each branch represents a different outcome of the measurement and the *corresponding* eigenstate for the [composite] state. All branches exist simultaneously in the superposition after any given sequence of observations.

Thus, in the case where an observation is made of the linear polarization state of a photon known to be initially in a state of circular polarization, the act of measurement causes the universe to split into two separate universes. In one of these universes, an observer measures and records that the photon was detected in a state of vertical polarization. In the other, the same observer measures and records that the photon was detected in a state of horizontal polarization. The observer now exists in two distinct states in the two universes. Looking back at the paradox of Schrödinger's cat, we can see that the difficulty is now resolved. The cat is not simultaneously alive and dead in one and the same universe, it is alive in one branch of the universe and dead in the other.

With repeated measurements, the universe, together with the observer, continues to split in the manner shown schematically in Fig. 5.7. In each

---

[†] Everett III, Hugh (1957). *Reviews of Modern Physics*. **29**, 454.

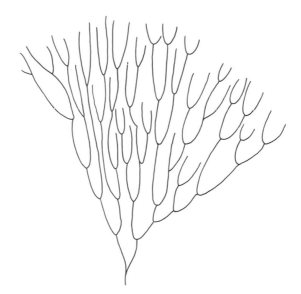

Fig. 5.7   Representation of a branching universe. A repeated measurement for which there are two possible outcomes continually splits the universe. The path followed from the beginning of the 'tree' to the end of one of its branches corresponds to a particular sequence of results observed in one of the split universes.

branch, the observer records a different sequence of results. Because each particular state of the observer does not perceive the universe to be branching, the results appear entirely consistent with the notion that the wavefunction of the circularly polarized photon collapsed into one or other of the two measurement eigenstates.

Why does the observer not retain some sensation that the universe splits into two branches at the moment of measurement? The answer given by the proponents of the Everett theory is that the laws of quantum mechanics simply do not allow the observer to make this kind of observation. DeWitt argued that if the splitting were to be observable, then it should be possible in principle to set up a second measuring device to obtain a result from the memory of the first device which differs from that obtained by its own direct observation. Wigner's friend could respond with an answer which differs from the one that Wigner could check for himself. This not only never happens (except where a genuine human error occurs) but is also not allowed by the mathematics. The branching of the universe is unobservable.

In a footnote added to the proof of his July 1957 paper, Everett accepted that the idea of a branching universe appears to contradict our everyday experience. However, he defended his position by noting that when Copernicus first suggested that the earth revolves around the sun (and not the other way around), this view was criticized on the grounds that the motion of the earth could not be felt by any of its inhabitants. In this case, our inability to sense the earth's motion was explained by Newtonian physics. Likewise, our inability to sense a splitting of the universe into different branches is explained by quantum physics.

'Schizophrenia with a vengeance'

If the act of measurement has no special place in the many-worlds interpretation, then there is no reason to define measurement to be distinct from any process involving a quantum transition between states. Now there have been a great many quantum transitions since the big bang origin of the universe, some 15 billion years ago. Each transition will have involved the development of a superposition of wavefunctions with each term in the superposition corresponding to different final states of the transition. Each transition will have therefore split the universe into as many branches as there were terms in the superposition. DeWitt has estimated that by now there must be more than 10^{100} branches.

Some of these branches will be almost indistinguishable from the one I (and presumably you) currently inhabit. Some will differ only in the way the polarized photons scattered from the surface of the VDU on which I am composing these words interact with the light-sensitive cells in my eyes. Many of these branches will contain almost identical copies of this book, being read by almost identical copies of you. No wonder DeWitt called the many-worlds interpretation 'schizophrenia with a vengeance'.

That there may exist 'out there' a huge number of different universes is a rather eerie prospect. Many of these universes will contain the same arrangement of galaxies that we can see in our 'own' universe. Some will contain a small, rather insignificant G2-type star identical to our own sun with a beautiful, but fragile-looking blue–green planet third from the centre in its planetary system. But in some of these branches the kinds of quantum transitions involving cosmic rays and giving rise to chance mutations in living creatures will have turned out differently from the ones which occurred in our earth's past history. Perhaps in some of these branches, mankind has not evolved and life on earth is dominated by a different species. An individual quantum transition may appear an unimportant event, but perhaps it can have ultimately profound consequences.

Parallel universes

When Everett presented his theory, he wrote of the observer state 'branching' into different states and drew an analogy with the branches of a tree. However, more recent variants of Everett's original interpretation have been proposed, in which the universe with which we are familiar is but one of a very large number (possibly an infinite number) of *parallel* universes. Thus, instead of the universe splitting into separate branches as a result of a quantum transition, the different terms of the superposition are partitioned between a number of already existing parallel universes.

Perhaps the major difference between this interpretation and the Everett original is that it allows for the possibility that the parallel universes may interact and *merge*. Indeed, it has been argued that we obtain indirect evidence for such a merging every time we perform an interference experiment. For example, we know that a single photon passes through a double slit apparatus and may be detected on the other side using a piece of photographic film. In the original Everett interpretation, we would say that the universe splits into two branches. In one of the branches, the photon passes through one of the slits whereas in the other branch it passes through the other slit. In the parallel universes interpretation, the wavefunction of the photon is partitioned between two universes as it passes through the slits, but these universes then merge once again to produce a single photon which is detected. In either universe there was a photon which followed a completely deterministic trajectory through one or other of the two slits, but the interaction of the universes produces an interference in which it is no longer possible to say which slit the photon went through.

This gives rise to another interesting possibility which has been pursued by the astrophysicist David Deutsch. Imagine that we set up a double slit experiment in which an observer determines which of the two slits the photon goes through. He agrees with us beforehand to note down that he definitely perceives the photon to go through one of the slits but he does not tell us which. The experiment is performed and the result enters the observer's memory. He writes in his notebook that he definitely saw the photon to pass through one of the slits. Now, according to the von Neumann–Wigner interpretation of the quantum measurement process, the wavefunction of the photon collapsed when it encountered the observer's consciousness. The observed result is therefore the *only* result and the other has 'disappeared' in the sense that its probability has been reduced to zero by the act of measurement. However, in the parallel universes interpretation, *both* results are obtained in two different universes and, in principle, an interference effect can still be observed if we can somehow merge them together. This would be equivalent to merging the quantum memory states of the two observers in the two universes to produce an interference. Thus, having noted that he saw the photon to be detected to pass through one of the slits, an observer whose memory states interfere 'forgets' which one it was.

The observer must feel very odd under these circumstances. He remembers that he saw the photon to be detected passing through one of the slits but cannot remember which one it was. This is in complete contrast to the situation obtained if the wavefunction collapses, since in this case the observer will remember which slit the photon went through. Here then is a proposal for a laboratory test of the parallel universes interpretation of quantum mechanics. Unfortunately, the brain does not appear to function at the level of individual quantum events. If it did, we would be able to 'feel' every quantum transition occurring inside our brains — not a very appealing prospect. However, Deutsch has suggested that it may one day be possible to construct an artificial brain capable of functioning at the quantum level. Instead of performing this experiment with a human observer, we would ask this artificial brain to perform the experiment for us, and simply ask it what it felt.

It has been said that the many-worlds interpretation of quantum theory is cheap on assumptions, but expensive with universes. That such a bizarre interpretation can result from the simplest of solutions to the quantum measurement problem demonstrates how profound the problem is. Furthermore, the difficulties raised by the kind of non-locality revealed by the Aspect experiments are not removed by Everett's theory, and we are still left with the need to invoke an instantaneous, faster-than-light splitting of or partitioning between universes. Although John Wheeler was an early champion of Everett's approach, he later rejected the theory 'because there's too much metaphysical baggage being carried along with it'. Until such time as Deutsch's artificial quantum brain can be constructed, our judgement of the many-worlds interpretation must remain a personal one.

5.5 THE HAND OF GOD?

Einstein's comment that 'God does not play dice' is one of the best known of his many remarks on quantum theory and its interpretation. Niels Bohr's response is somewhat less well known: 'But still, it cannot be for us to tell God, how he is to run the world.'

Is it possible that after centuries of philosophical speculation and scientific research on the nature of the physical world we have, in quantum theory, finally run up against nature's grand architect? Is it possible that the fundamental problems of interpretation posed by quantum theory in its present form arise from our inability to fathom the mind of God? Are we missing the 'ultimate' hidden variable? Could it be that behind every apparently indeterministic quantum measurement we can discern God's guiding hand?

Away from the cut-and-thrust of their scientific research papers, Einstein, Bohr and their contemporaries spoke and wrote freely about God and his designs. To a limited extent, this habit continues with modern-day scientists. For example, in *A brief history of time*, Stephen Hawking writes in a relaxed way about a possible role for God in the creation of the universe and in *The emperor's new mind*, Roger Penrose writes of 'God-given' mathematical truth. Among scientists, speculating about God in the open literature appears to be the preserve of those who have already established their international reputations. We would surely raise our eyebrows on discovering that the research programme of a young, struggling academic scientist in the 1990s is organized around his desire to know how God created the world. We would at least anticipate that such a scientist may have difficulties securing the necessary funding to carry out his research.

But discovering more about how God created the world was all the motivation Einstein needed for his work. Admittedly, Einstein's was not the traditional medieval God of Judaism or Christianity, but an impersonal God *identical* with Nature: *Deus sive Natura* — God or Nature — as described by the seventeenth century philosopher Baruch Spinoza.

And herein lies the difficulty. In modern times it is almost impossible to resist the temptation to equate belief in God with an adherence to a religious philosophy or orthodoxy. Scientists are certainly taught not to allow their scientific judgement to be clouded by their personal beliefs. Religious belief entails blind acceptance of so many dogmatic 'truths' that it negates any attempt at detached, rational, scientific analysis. In saying this, I do not wish to downplay the extremely important sociological role that religion plays in providing comfort and identity in an often harsh and brutal world. But once we accept God without religion, we can ask ourselves the all-important questions with something approaching intellectual rigour. The fact that we have lost the habit or the need to invoke the existence of God should not prevent us from examining this possibility as a serious alternative to the interpretations of quantum theory discussed previously. It is, after all, no less metaphysical or bizarre than some of the other possibilities we have considered so far.

Does God exist?

There is a timelessness about this question. It has teased the intellects of philosophers for centuries and weaves its way through the entire history of philosophical thought. Even in periods where it may have been generally accepted to be a non-question, it has lurked in the shadows, biding its time.

For many centuries philosophical speculation regarding the existence of God was so closely allied to theology as to be essentially indistinguishable from it. In the thirteenth century Thomas Aquinas helped to restore Aristotelian philosophy and science, an ancient learning that had been buried and all but forgotten during the 'Dark Ages'. But Aquinas was a scholar of the Roman Catholic Church, and he took great pains to ensure that pagan and other heretical elements were carefully weeded out of Aristotle's philosophy. The Church elevated Aristotelianism to the exalted status of a religious dogma and so pronounced on all matters not only of religious faith, but also of science. To contradict the accepted wisdom of the Church was to court disaster, as Galileo discovered on the 22 June 1633, when at the end of his trial for heresy he was forced to abjure the Copernican doctrine and was placed under house arrest.

Against this background, a seventeenth century philosopher wishing to establish a new philosophical tradition had to tread warily. René Descartes had just completed his magnum opus, which he had called *De mundo*, when in November 1633 he received news of Galileo's trial and condemnation. Descartes was dismayed: the Copernican system formed the basis of his work, and it became clear that if he published it, it would not have the effect he had hoped. Instead, he chose to 'leak' bits of it out, hoping always to stay on the right side of the Church authorities. It is perhaps not surprising that when he published his *Meditations*, in which he offered three different proofs for the existence of God, he decided to dedicate this work to the Dean and Doctors of the Sacred Faculty of Theology of Paris.

Descartes's aim was not to subvert the teachings of the Church, but to demonstrate that the orthodox conclusions regarding the soul of man and the existence of God could be reached using the power of reason. His intention was to bring something approaching mathematical rigour to bear on these philosophical questions. Having said that, it is apparent that Descartes's arguments fall somewhat short of the ideal which he had set for himself in his *Discourse on method*, published four years earlier. Nevertheless, his approach marked a distinct break with the past.

The ontological proof

Descartes advanced three proofs for the existence of God. Two are to be found in his 'Third meditation', but the one from which he seemed to derive most pleasure is found in his 'Fifth meditation'. This is the so-called ontological proof or ontological argument.

Remember that Descartes had already established (with certainty) that he is a thinking being and that, therefore, he exists. As a thinking being, he recognizes that he is imperfect in many ways, but he can conceive of the *idea* of a supremely perfect being, possessing all possible perfections. Now it goes without saying that a being that is imperfect in any way is not a supremely perfect being. Descartes assumed that existence is a perfection, in the sense that a being that does not exist is imperfect. Therefore, he reasoned, it is self-contradictory to conceive of God as a supremely perfect being that does not exist and so lacks a perfection. Such a notion is as absurd as trying to conceive of a triangle that has only two angles. Thus, God must be conceived as a being who exists. Hence, God must exist.

If you were expecting some dramatic revelation from this argument, you will have probably been disappointed. But then, we should remember the circumstances and influences under which Descartes deduced this proof. He could not have come up with any other answer because he believed God to exist. All he wanted to do was to establish this as a fundamental truth through pure reason. His greatest contribution to philosophy is not to be found in the answers he produced to philosophical questions, but in the methods he used to arrive at them.

The cosmological proof

The methods used by Descartes were picked up by other philosophers of his time, although many did not always feel it necessary to indulge in the kind of systematic doubting that Descartes had thought to be important. Thus, the German philosopher Gottfried Wilhelm Leibniz was happy to accept as self-evident much of what Descartes had taken great pains to prove, and adapted and extended many other elements in Descartes's line of reasoning. For example, in developing his own philosophical position, Leibniz was happy to accept the existence of the world also to be self-evident, although its nature might not be.

Like his predecessor, Leibniz also presented three proofs for the existence of God. Two of these are similar to two of Descartes's proofs. The third, which is usually known as the cosmological proof or the cosmological argument, was published in 1697 in Leibniz's essay *On the ultimate origination of things*.

Leibniz's argument is based on the so-called principle of *sufficient reason*, which he interpreted to mean that if something exists, then there must be a good reason. Thus, the existence of the world and of the eternal truths of mathematics and logic must have a reason. Something must have caused these things to come into existence. He claimed that there is within the world itself no sufficient reason for its own existence. As time elapses, the state of the world evolves according to certain physical laws of change. It could be argued, then, that the cause of the existence of the world at any one moment is to be found in the existence of the world just a moment before. Leibniz rejected this argument:[†] '. . . however far you go back to earlier states, you will never find in those states a full reason why there should be any world rather than none, and why it should be such as it is.'

The world cannot just *happen* to exist, and whatever (or whoever) caused it to exist must also exist, since the principle of sufficient reason demands that something cannot come from nothing: *ex nihilo, nihilo fit*. Furthermore, the ultimate, or first, cause of the world must exist outside the world. Of course, this first cause is God. God is the only sufficient reason for the existence of the world. The world exists, therefore it is necessary for God also to exist.

The cosmological proof has a long history. Plato used something akin to it in his discussion of God-as-creator in the *Timaeus*. It also has an entirely modern applicability. We now have good reason to believe that the world (which in its modern context we take to mean the universe) was formed about 15 billion years ago in the big bang space–time singularity. The subsequent expansion of space–time has produced the universe as we know it today, complete with galaxies, stars, planets and living creatures. Modern theories of physics and chemistry allow us to deduce the reasons for the existence of all these things (possibly including life) based on the earlier states of the universe. In other words, once the universe was off to a good start, the rest followed from fundamental physical and chemical laws. Scientists are generally disinclined to suggest that we need to call on God to explain the evolution of the post big bang universe. But the universe had a *beginning*; which implies that it must have had a first cause. Do we need to call on God to explain the big bang? Stephen Hawking writes:[‡] 'An expanding universe does not preclude a creator, but it does place limits on when he might have carried out his job!'

To be sure, there are a number of theories that suggest the big bang might not have been the beginning of the universe but only the beginning of the present phase of the universe. These theories invoke endless cycles each consisting of a big bang, expansion, contraction and collapse of the universe in a 'big crunch', followed by another bang and expansion. It

[†] Leibniz, Gottfried Wilhelm (1973). *Philosophical writings*. J.M. Dent, London.
[‡] Hawking, Stephen W. (1988). *A brief history of time*. Bantam Press, London.

has even been suggested that the laws of physics might be redefined at the beginning of every cycle. However, this does not solve the problem. In fact, we come right back to Leibniz's argument about previous states of the world not providing sufficient reason for the existence of the current state of the world.

God or Nature

Although Baruch Spinoza was a contemporary of Leibniz, his views concerning God could not have been more different. The work of Spinoza represents a radical departure from the pseudo-religious conceptions of God advanced by both Descartes and Leibniz. A Dutch Jew of Portugese descent living in a largely Christian society, Spinoza was ostracized by both the Jewish and Christian communities as an atheist and a heretic. This isolation suited his purposes well, since he wished to work quietly and independently, free of more 'earthly' distractions. It is not that Spinoza did not believe there to be a God, but his reasoning led him to the conclusion that God is *identical* with nature rather than its external creator.

Spinoza's argument is actually based on his ideas regarding the nature of substance. He distinguished between substances that could exist independently of other things and those that could not. The former substances provide in themselves sufficient reason for their existence — they are their own causes (*causa sui*) — and no two substances can possess the same essential attributes. He then defined God to be a substance with infinite attributes. Since different substances cannot possess the same set of attributes, it follows logically that if a substance with infinite attributes exists then this must be the *only* substance that can exist: 'Whatever is, is in God.'

Spinoza's seventeenth century conception of God is quite consistent with twentieth century thinking. His is not the omniscient, omnipresent God of Judao-Christian tradition, who is frequently imagined to be an all-powerful being with many human-like attributes (such as mind and will). Rather, Spinoza's God is the embodiment of everything in nature. The argument is that when we look at the stars, or on the fragile earth and its inhabitants, we are seeing the physical manifestations of the attributes of God. God is not outside nature — he did not shape the fundamental physical laws by which the universe is governed — he *is* nature. Neither is he a free agent in the sense that he can exercise a freedom of will outside fundamental physical laws. He is free in the sense that he does not rely on an external substance or being for his existence (he is *causa sui*). He is a deterministic God in that his actions are determined by his nature.

This is the kind of God with which most western scientists would feel reasonably comfortable, if they had to accept that a God exists at all. The fact that modern physics has been so enormously successful in defining the character of physical law does not reduce the power of the argument in favour of the existence of Spinoza's God. Indeed, in his book *God and the new physics*, Paul Davies suggests that science 'offers a surer path to God than religion'.[†] Although scientists tend not to refer in their papers to God as such, with the advent of modern cosmology and quantum theory, some have argued that the need to invoke a 'substance with infinite attributes' is more compelling than ever.

As mentioned earlier, Einstein's frequent references to God were references to Spinoza's God. In his studies, he was therefore concerned to discover more about 'God or Nature'. This does not mean to say that Einstein did not believe that there must be some kind of divine plan or order to the universe. This much is obvious from his adherence to strict causality and determinism and his later opposition to the Copenhagen interpretation. He expected to find *reason* in nature, not the apparent trusting to luck suggested by quantum indeterminism.

A world without God

The triumphs of seventeenth century science clearly demonstrated that the Aristotelian dogma espoused by the Church was completely untenable. As the grip of the Church relaxed and public opinion became generally more liberal, so it became possible for a final parting of the ways between philosophy and theology. This transition was achieved by two giants of eighteenth century philosophy—David Hume and Immanuel Kant.

Hume demolished both the ontological and cosmological proofs for the existence of God in his *Dialogues concerning natural religion*. That this work was still controversial is evidenced by the fact the Hume preferred to arrange for its publication after his death (it was published in 1779). Some sense of Hume's situation can be gleaned from a quotation which appeared on the title page of his massive work *A treatise of human nature*, published in 1739: 'Seldom are men blessed with times in which thay may think what they like, and say what they think.'[‡]

Most of Hume's arguments, which are made through the agency of a dialogue between three fictional characters, hinge around the contention that there is an inherent limit to what can be rationally claimed through metaphysical speculation and pure reason. He presents the case that the

[†] Davies, Paul (1984). *God and the new physics*. Penguin, London.
[‡] Hume, David (1969). *A treatise of human nature*. Penguin, London.

earlier conclusions regarding the existence of God made by Descartes and Leibniz and others simply do not stand up to close scrutiny. They fail because too many assumptions are made without justification. Why should existence be regarded as a perfection, as Descartes assumed? Why is it necessary for the world to have a cause, whereas God does not (indeed, cannot). Why not simply conclude that the world itself needs no cause, eliminating the need for God? Surely the wretched state of mankind is itself sufficient evidence that the benevolent God of Christian tradition *cannot* exist?

Although Kant, coming a few years after Hume, did not entirely accept Hume's outright rejection of metaphysics, the die was effectively cast. Kant's *Critique of pure reason*, published in 1781, picked up more-or-less where Hume left off. In this work, Kant concluded that all metaphysical speculation about God, the soul and the natures of things cannot provide a path to knowledge. True knowledge can be gained only through experience and, since we appear to have no direct experience of God as a supreme being, we are not justified in claiming that he exists. However, unlike Hume, it was not Kant's intention to develop a purely empiricist philosophy, in which all things that we cannot know through experience are rejected. We must *think* of certain things as existing in themselves even though we cannot know their precise natures from the ways in which they appear to us. Otherwise, we would find ourselves concluding that an object can have an appearance without existence, which Kant argued to be obviously absurd.

Thus, Kant did not reject metaphysics *per se*, but redefined it and placed clear limits on the kind of knowledge to be gained through speculative reasoning. There is still room for religion in Kant's philosophy, and he argues that there are compelling practical reasons why *faith*, as distinct from knowledge, is important: 'I must, therefore, abolish *knowledge* to make room for *belief*'[†], meaning that belief in God and the soul of man is not founded on knowledge of these things gained through speculative reason, but requires an act of faith. This does not have to be religious faith in the usual sense; it can be a very practical faith which is necessary to make the connection between things as they appear and the things-in-themselves of which we can have no direct experience.

Like Hume, Kant also demolished the ontological and cosmological proofs for the existence of God, because these arguments necessarily transcend experience. Thus, any attempt to prove the existence of God requires assumptions that go beyond our conscious experience and cannot therefore be justified. Belief in the existence of God is not something

[†] Kant, Immanuel (1934). *Critique of pure reason*, (trans. J. M. D. Meiklejohn). J. M. Dent, London.

that can be justified by pure reason, but may be justified through faith. This does not make God unnecessary, but it does limit what we can know of him.

The fundamental shift in the direction of philosophical thought which was initiated by Hume and Kant was continued and reinforced by philosophers in the nineteenth century. The divorce of philosophy from religion became permanent. Hume's outright rejection of metaphysical speculation as meaningless was eventually to provide one of the inspirations for the Vienna Circle. Indeed, logical positivism represents the ultimate development of the kind of empiricism advocated by Hume. As we discussed in Section 3.1, the positivist philosophy is based on what we can say meaningfully about what we experience. With the positivists of the twentieth century, philosophy essentially became an *analytical* science. Wittgenstein once remarked that the sole remaining task for philosophy is the analysis of language.

Quantum physics and metaphysics

Despite the positivists' efforts to eradicate metaphysics from philosophy, the old metaphysical questions escaped virtually unscathed. I find it rather fascinating to observe that although the possibility of the existence of God and the relationship between mind and body no longer form part of the staple diet of the modern philosopher of science, they have become increasingly relevant to discussions on modern quantum physics. Three centuries of gloriously successful physics have brought us right back to the kind of speculation that it took three centuries of philosophy to reject as meaningless. It may be that the return to metaphysics is really a grasping at straws — an attempt to provide a more 'acceptable' world view until such time as the further subtleties of nature can be revealed in laboratory experiments and this agonizing over interpretation thereby relieved. But we have no guarantee that these subtleties will be any less bizarre than quantum physics as it stands at present.

And what of God? Does quantum theory provide any support for the idea that God is behind it all? This is, of course, a question that cannot be answered here, and I am sure that readers are not expecting me to try. Like all of the other possible interpretations of quantum theory discussed in this chapter, the God-hypothesis has many things to commend it, but we really have no means (at present) by which to reach a logical, rational preference for any one interpretation over the others. If some readers draw comfort from the idea that either Spinoza's God or God in the more traditional religious sense presides over the apparent uncertainty of the quantum world, then that is matter for their own personal faith.

Closing remarks

We have now come to the end of our guided tour of the meaning of quantum theory: I hope you enjoyed it. I have tried to be an impartial guide in the sense that I have tried not to argue from a particular position. In fact, I hope that I have argued for all the different positions described in this book with something approaching equal force. This has been necessary to capture the lively nature of a debate which has been going on for over 60 years. It has been necessary, moreover, to get across the important message that quantum theory has more than one interpretation.

It is usual at the end of a tour such as this one for the guide to be asked *his* opinion. I have read a number of books written recently by physicists in which all the experimental evidence against the notion of local reality has been carefully weighed, but which then close with some kind of final plea for an independent reality. I hope I have done enough in this book to demonstrate that, no matter where we start, we always return to the central philosophical arguments of the positivist versus the realist. The conflict between these philosophical positions formed the basis of the Bohr–Einstein debate. No matter what the state of experimental science, the conflict between the positivists' conception of an empirical reality and the realists' conception of an independent reality can *never* be resolved. The experimental results described in Chapter 4 cannot shake the realists' deeply felt belief in an independent reality, although they certainly make it a more complicated reality than might at first have been thought necessary. Thus, any final plea for an independent reality is really an appeal to faith, in the sense that the realist must ultimately accept the logic of the positivists' argument but will still not be persuaded.

To some extent, I myself am not deeply troubled by the prospect of a reality which is not independent of the observer or the measuring device. However, I do not share the uncompromising views characteristic of the positivist. I am convinced that the desire to relate their theories to elements of an independent reality is part of the psychological make-up of many scientists. They feel it is necessary to try continually to go beyond the symbols in a mathematical equation and attach a deeper meaning to them. Without this continual attempt to penetrate to an underlying reality, science would be a sterile, passive and rather unemotional activity. This it certainly is not. Like all acts of faith, the search for an independent reality involves striving for a goal that can never be reached. This does not mean that the effort is any less worthwhile. On the contrary, it is through this process of striving for the unachievable that progress in science is made.

With regard to quantum theory, my personal view is that we still do not yet know enough about the physical world to make a sound judgement about its meaning. The positivist says that the theory is all there is, but the realist says: Look again, we do not yet have the whole story. As to where we might look, my recommendation is to watch *time* closely: we do not yet seem to have a good explanation of it. This is not to say that a better understanding of time will automatically solve all the conceptual problems of quantum theory. Time, I suppose, will tell.

I am reasonably certain of one thing. The unquestioning acceptance of the Copenhagen interpretation of quantum theory has, in the last 40 years or so, held back progress on the development of alternative theories. It has been very difficult for the voices raised against the orthodox interpretation to be heard. Remember that it was John Bell — an opponent of the dogmatic Copenhagen view — whose curiosity and determination led to Bell's theorem and ultimately to new experimental tests. Blind acceptance of the orthodox position cannot produce the challenges needed to push the theory eventually to its breaking point. And break it will, probably in a way no-one can predict to produce a theory no-one can imagine. The arguments about reality will undoubtedly persist, but at least we will have a better theory.

I have tried to argue that quantum theory is a difficult subject for the modern undergraduate student of physical science because its interpretation is so firmly rooted in philosophy. If, in arguing this case, I have only made the subject seem even more confusing, then I apologize. However, my most important message is a relatively simple one: quantum theory is rife with conceptual problems and contradictions. If you find the theory difficult to understand, this is the theory's fault — not yours.

Appendix A
Planck's derivation of the radiation law

Max Planck had struggled with the theory of black-body radiation for about six years before the end of the nineteenth century. In 1897, he used a model of simple, so-called Hertzian oscillators to calculate how $\rho(\nu, T)$ should depend on the mean internal energy U of an individual oscillator. This is essentially a problem involving the interaction between a linear oscillator (with a certain mass and electric charge) and a monochromatic (single ν) electric field. As discussed in Section 1.1, it is not immediately clear just how these oscillators should be interpreted. However, with hindsight, we can see that the oscillators have many of the properties we would now associate with the atoms which constitute the material of a radiation cavity.

Planck's methods can be found in many physics textbooks, and so we will start here with his result:

$$\rho(\nu, T) = \frac{8\pi\nu^2}{c^3} U. \tag{A.1}$$

The Rayleigh–Jeans law can be obtained from this expression simply by setting $U = kT$, a step which can made by assuming a Maxwell–Boltzmann distribution of energy among the oscillators, but a step which Planck himself did not take.

Exploiting the analogy between radiation trapped in a cavity and a gas consisting of freely moving particles trapped in a container, Planck went on to determine the entropy associated with the Hertzian oscillators. For closed systems with constant volume that can do no work of expansion or compression, the first law of thermodynamics can be written in differential form as:

$$dU = T\mathrm{d}S \tag{A.2}$$

where S is the entropy. It follows that

$$\left(\frac{\partial S}{\partial U}\right)_V = \frac{1}{T} \tag{A.3}$$

where the subscript V indicates that the volume of the system is treated as a constant. This standard thermodynamic expression provides a connection between the entropy of the oscillators and Planck's radiation formula. From eqn (A.1) and the radiation formula, eqn (1.3), we have

$$U = \frac{h\nu}{e^{h\nu/kT} - 1} \tag{A.4}$$

which can be rearranged to give an expression for $1/T$ in terms of U:

$$\frac{1}{T} = \frac{k}{h\nu} \ln \left(\frac{h\nu}{U} + 1 \right) \tag{A.5}$$

or

$$\frac{1}{T} = \frac{k}{h\nu} \left[\ln \left(1 + \frac{U}{h\nu} \right) - \ln \frac{U}{h\nu} \right]. \tag{A.6}$$

Hence, from eqn (A.3),

$$dS = \int \frac{k}{h\nu} \left[\ln \left(1 + \frac{U}{h\nu} \right) - \ln \frac{U}{h\nu} \right] dU \tag{A.7}$$

where the integration is carried out for constant V. The result is:

$$S = k \left[\left(1 + \frac{U}{h\nu} \right) \ln \left(1 + \frac{U}{h\nu} \right) - \frac{U}{h\nu} \ln \frac{U}{h\nu} \right]. \tag{1.4}$$

This is easily verified by differentiating eqn (1.4) with respect to U. As discussed in Section 1.1, eqn (1.4) is an expression for the entropy of an oscillator which is consistent with Planck's radiation formula and therefore consistent with experiment. What is needed now is somehow to derive eqn (1.4) from the intrinsic properties of the oscillators themselves.

In 1877, Boltzmann proposed that a gas could be thought to consist of N distinguishable molecules. Each molecule was assigned a kinetic energy of 0, ε, 2ε, 3ε, . . ., $P\varepsilon$, where ε is an arbitrary unit of energy and P is an integer. The state of the gas could be specified by its 'complexion', i.e. by assigning each individual molecule a specific energy content (molecule 1 has energy 7ε, molecule 2 has energy 2ε, etc). The energy distribution is determined by the numbers of molecules with given energies (4 molecules with energy ε, 10 molecules with energy 2ε, etc). Thus, many complexions can have the same energy distribution. Boltzmann assumed that all complexions are equally likely, and calculated the most probable energy distribution W_N. He found that the entropy of the gas at equilibrium is directly related to $\ln W_N$. In fact, for an ensemble of N molecules, the entropy S_N is given by

$$S_N = k \ln W_N \qquad (A.8)$$

although this is a result that Boltzmann himself never stated.

In applying Boltzmann's ideas to the problem of black-body radiation, Planck had to assume that the total energy could be split up into P indistinguishable but independent elements, each with an energy ε, which are distributed over N distinguishable radiation oscillators. The number of ways of doing this is given by

$$W_N = \frac{(N - 1 + P)!}{(N - 1)! P!} . \qquad (A.9)$$

For example, if we have to dispose of four energy elements over two oscillators ($P = 4$, $N = 2$), then according to eqn (A.9), $W_N = 5$. These different distributions correspond to putting all four elements into one oscillator and none in the other (4ε, 0), three in one oscillator and one in the other (3ε, ε), (2ε, 2ε), (ε, 3ε) and (0, 4ε).

For all practical purposes, N and P are very large numbers and so W_N can be approximated as:

$$W_N = \frac{(N + P)!}{N! P!} . \qquad (A.10)$$

Furthermore, the factorials of very large numbers can be approximated using Stirling's formula, $N! = (N/e)^N$, giving

$$W_N = \frac{(N + P)^{(N+P)}}{N^N P^P} . \qquad (A.11)$$

An expression for the total entropy of the N oscillators can then be found by combining eqns (A.8) and (A.11):

$$S_N = k \left[(N + P) \ln (N + P) - N \ln N - P \ln P \right]. \qquad (A.12)$$

The total internal energy of the N oscillators is simply N times the mean internal energy of one oscillator, U. This same quantity (the total energy) is also given by the number of energy elements, P, multiplied by the size of each energy element ε, i.e. $NU = P\varepsilon$, and so

$$P = \frac{NU}{\varepsilon} . \qquad (A.13)$$

Inserting this expression for P into eqn (A.12) gives

$$S_N = kN \left[\left(1 + \frac{U}{\varepsilon} \right) \ln \left(1 + \frac{U}{\varepsilon} \right) - \frac{U}{\varepsilon} \ln \frac{U}{\varepsilon} \right] \qquad (A.14)$$

(note that all terms in $\ln N$ which appear in the resulting expression cancel). Thus, the entropy of an individual oscillator $(S = S_N/N)$ is given by:

$$S = k \left[\left(1 + \frac{U}{\varepsilon}\right) \ln \left(1 + \frac{U}{\varepsilon}\right) - \frac{U}{\varepsilon} \ln \frac{U}{\varepsilon} \right] \qquad (1.5)$$

which is the result quoted in Section 1.1. To obtain eqn (1.4) from eqn (1.5), it was necessary for Planck to assume that the energy elements are given by $\varepsilon = h\nu$. Thus, the radiation energy is not exchanged between the oscillators and the electromagnetic field continuously, but rather in discrete packets which Planck later called quanta.

Appendix B
Bell's inequality for non–ideal cases

The generalization of Bell's inequality requires that we consider experiments involving four different orientations of the two polarization analysers. We denote these orientations as *a, b, c* and *d*. We suppose that the results of measurements made on photon A and photon B are determined by some local hidden variable (or variables) denoted λ. The λ values are distributed among the photons according to a distribution function $\rho(\lambda)$, which is essentially the ratio of the number of photons with the value λ, N_λ, divided by the total number of photons. We assume that this function is suitably normalized, so that $\int \rho(\lambda)d\lambda = 1$.

The average or expectation values of the results of measurements made on individual photons depend on the particular orientation of the polarization analyser and the λ value. We denote the expectation value for photon A entering PA_1 set up with orientation *a* as $A(a, \lambda)$. Similarly, the expectation value for photon B entering PA_2 set up with orientation *b* is $B(b, \lambda)$. The possible result of each measurement is ± 1, corresponding to detection in the vertical or horizontal channels respectively. It must then follow that the absolute values of the expectation values cannot exceed unity, i.e.

$$|A(a, \lambda)| \leqslant 1, \quad |B(b, \lambda)| \leqslant 1. \qquad (B.1)$$

We assume that the individual results for A depend on *a* and on λ, but are independent of *b* and vice versa (Einstein separability).

The expectation value for the joint measurement of A and B, $E(a, b, \lambda)$ is given by the product $A(a, \lambda)B(b, \lambda)$. We can eliminate λ from this expression by averaging the results over many photon pairs (emphasizing the statistical nature of the hidden variable approach), or by integrating over all λ:

$$E(a, b) = \int A(a, \lambda) B(b, \lambda) \rho(\lambda)d\lambda. \qquad (B.2)$$

This follows if we assume that we can perform measurements on a sufficiently large number of photon pairs so that all possible values of λ are sampled. Much the same kind of reasoning can be used to show that

$$E(a, b) - E(a, d) = \int [A(a, \lambda)B(b, \lambda) - A(a, \lambda)B(d, \lambda)]\rho(\lambda)d\lambda$$

$$= \int A(a, \lambda)[B(b, \lambda) - B(d, \lambda)]\rho(\lambda)d\lambda \qquad \text{(B.3)}$$

and so, since $|A(a, \lambda)| \leqslant 1$,

$$|E(a, b) - E(a, d)| \leqslant \int |B(b, \lambda) - B(d, \lambda)|\rho(\lambda)d\lambda. \qquad \text{(B.4)}$$

Similarly,

$$|E(c, b) + E(c, d)| \leqslant \int |B(b, \lambda) + B(d, \lambda)|\rho(\lambda)d\lambda. \qquad \text{(B.5)}$$

Combining eqns (B.4) and (B.5) gives

$$|E(a, b) - E(a, d)| + |E(c, b) + E(c, d)| \qquad \text{(B.6)}$$

$$\leqslant \int [|B(b, \lambda) - B(d, \lambda)| + |B(b, \lambda) + B(d, \lambda)|]\rho(\lambda)d\lambda.$$

From (B.1), it is apparent that $|B(b, \lambda) - B(d, \lambda)| + |B(b, \lambda) + B(d, \lambda)|$ must be less than or equal to 2. Thus,

$$|E(a, b) - E(a, d)| + |E(c, b) + E(c, d)| \leqslant 2\int \rho(\lambda)d\lambda \qquad \text{(B.7)}$$

and, since by definition $\int \rho(\lambda)d\lambda = 1$,

$$|E(a, b) - E(a, d)| + |E(c, b) + E(c, d)| \leqslant 2. \qquad \text{(4.37)}$$

Note that nowhere in this derivation have we needed to assume that we will obtain perfect correlation between the measured results for any combination of the analyser orientations. Equation (4.37) is therefore valid for non-ideal cases in which limitations in the experimental apparatus prevent the observation of perfect correlation.

Bibliography

ADVANCED TEXTS

Bohm, David (1951). *Quantum theory*. Prentice-Hall, Englewood Cliffs, NJ.

Bohm, David (1980). *Wholeness and the implicate order*. Routledge, London.

Davies, P.C.W. (1974). *The physics of time asymmetry*. Surrey University Press.

Dirac, P.A.M. (1958). *The principles of quantum mechanics*, (4th edn.). Clarendon Press, Oxford.

d'Espagnat, Bernard (1989). *The conceptual foundations of quantum mechanics*, (2nd edn.). Addison-Wesley, New York.

Hermann, Armin (1971). *The genesis of quantum theory (1899–1913)*. MIT Press, Cambridge, MA.

Jammer, Max (1974). *The philosophy of quantum mechanics*. Wiley, New York.

Merzbacher, Eugene (1970). *Quantum mechanics*, (2nd edn.). Wiley, New York.

Prigogine, Ilya (1980). *From being to becoming*. W.H. Freeman, San Francisco, CA.

von Neumann, John (1955). *Mathematical foundations of quantum mechanics*. Princeton University Press.

UNDERGRADUATE TEXTS

Atkins, P.W. (1983). *Molecular quantum mechanics*, (2nd edn.). Oxford University Press.

Dodd, J.E. (1984). *The ideas of particle physics*. Cambridge University Press.

Feynman, Richard P., Leighton, Robert B., and Sands, Matthew (1965). *The Feynman lectures on physics*, Vol. III. Addison-Wesley, Reading MA.

French, A.P. (1968). *Special relativity*. Van Nostrand Reinhold, Wokingham.

French, A.P. and Taylor, E.F. (1978). *An introduction to quantum physics*. Van Nostrand Reinhold, Wokingham.

Levine, Ira N. (1983). *Quantum chemistry*, (3rd edn.). Allyn & Bacon, Boston, MA.

Rae, Alastair I.M. (1986). *Quantum mechanics*, (2nd edn.). Adam Hilger, Bristol.

Rindler, Wolfgang (1982). *Introduction to special relativity*. Oxford University Press.

BIOGRAPHIES

Bernstein, Jeremy (1991). *Quantum profiles*. Princeton University Press. (Biographical sketches of John Bell and John Wheeler).

Cassidy, David C. (1992). *Uncertainty: the life and science of Werner Heisenberg*. W.H. Freeman, New York.

Feynman, Richard P. (1985). *'Surely you're joking, Mr. Feynman!'*. Unwin, London.

Hoffmann, Banesh (1975). *Albert Einstein*. Paladin, St. Albans.

Klein, Martin J. (1985). *Paul Ehrenfest: the making of a theoretical physicist*, Vol. 1, (3rd edn.). North-Holland, Amsterdam.

Kragh, Helge S. (1990). *Dirac: a scientific biography*. Cambridge University Press.

Moore, Walter (1989). *Schrödinger: life and thought*. Cambridge University Press.

Pais, Abraham (1982). *Subtle is the Lord: the science and the life of Albert Einstein*. Oxford University Press.

Popper, Karl (1976). *Unended quest: an intellectual autobiography*. Fontana, London.

ANTHOLOGIES AND COLLECTIONS

Bell, J.S. (1987). *Speakable and unspeakable in quantum mechanics*. Cambridge University Press.

Born, Max (1969). *Physics in my generation*, (2nd edn.). Springer, New York.

DeWitt, B.S. and Graham, N. (eds.) (1975). *The many worlds interpretation of quantum mechanics*. Pergamon, Oxford.

French, A.P. and Kennedy, P.J. (eds.) (1985). *Niels Bohr: a centenary volume*. Harvard University Press, Cambridge, MA.

Heisenberg, Werner (1983). *Encounters with Einstein*. Princeton University Press.

Hiley, B.J. and Peat, F.D. (eds.) (1987). *Quantum implications*. Routledge & Kegan Paul, London.

Kilmister, C.W. (ed.) (1987). *Schrödinger: centenary celebration of a polymath*. Cambridge University Press.

Schilpp, P.A. (ed.) (1949). *Albert Einstein: philosopher-scientist*. The Library of Living Philosophers, Open Court, La Salle, IL.

Wheeler, John Archibald and Zurek, Wojciech Hubert (eds.) (1983). *Quantum theory and measurement*. Princeton University Press.

PHILOSOPHY

Ayer, A.J. (1936). *Language, truth and logic*. Penguin, London.

Ayer, A.J. (1956). *The problem of knowledge*. Penguin, London.

Ayer, A. J. (1976). *The central questions of philosophy*. Penguin, London.

Descartes, René (1968). *Discourse on method and the meditations*. Penguin, London.

d'Espagnat, Bernard (1989). *Reality and the physicist*. Cambridge University Press.

Harré, R. (1972). *The philosophies of science*. Oxford University Press.

Heisenberg, Werner (1989). *Physics and philosophy*. Penguin, London.

Hume, David (1948). *Dialogues concerning natural religion*. Hafner Press, New York.

Hume, David (1969). *A treatise of human nature*. Penguin, London.

Jeans, James (1981). *Physics and philosophy*. Dover, New York.

Kant, Immanuel (1934). *Critique of pure reason*, (trans. J.M.D. Meiklejohn). J.M. Dent, London.

Kuhn, Thomas S. (1970). *The structure of scientific revolutions*, (2nd edn.). University of Chicago Press.

Leibniz, Gottfried Wilhelm (1973). *Philosophical writings*. J.M. Dent, London.

Murdoch, Dugald (1987). *Niels Bohr's philosophy of physics*. Cambridge University Press.

Popper, Karl R. (1959). *The logic of scientific discovery*. Hutchinson, London.

Popper, Karl R. (1982). *Quantum theory and the schism in physics*. Unwin Hyman, London.

Popper, Karl R. (1990). *A world of propensities*. Thoemmes, Bristol.

Russell, Bertrand (1967). *The problems of philosophy*. Oxford University Press.

Ryle, Gilbert (1963). *The concept of mind*. Penguin, London.

Schacht, Richard (1984). *Classical modern philosophers*. Routledge & Kegan Paul, London.

Scruton, Roger (1986). *Spinoza*. Oxford University Press.

POPULAR SCIENCE

Davies, Paul (1984). *God and the new physics*. Penguin, London.

Davies, Paul (1988). *Other worlds*. Penguin, London.

Davies, P.C.W. and Brown, J.R. (eds.) (1986). *The ghost in the atom*. Cambridge University Press.

Feynman, Richard (1967). *The character of physical law*. M.I.T. Press, Cambridge, MA.

Gamow, George (1965). *Mr Tompkins in paperback*. Cambridge University Press.

Gleick, James (1988). *Chaos: making a new science*. Heinemann, London.

Gregory, Bruce (1988). *Inventing reality: physics as language*. John Wiley & Sons, New York.

Hawking, Stephen W. (1988). *A brief history of time*. Bantam, London.

Koestler, Arthur (1964). *The sleepwalkers*. Penguin, London.

Lockwood, Michael (1990). *Mind, brain and the quantum: the compound 'I'*. Blackwell, Oxford.

Penrose, Roger (1990). *The emperor's new mind*. Vintage, London.
Polkinghorne, J.C. (1984). *The quantum world*. Penguin, London.
Prigogine, Ilya and Stengers, Isabelle (1985). *Order out of chaos*. Fontana, London.
Rae, Alastair (1986). *Quantum physics: illusion or reality?* Cambridge University Press.
Rohrlich, Fritz (1987). *From paradox to reality*. Cambridge University Press.
Sachs, Mendel (1988). *Einstein versus Bohr: the continuing controversies in physics*. Open Court, La Salle, IL.
Snow, C.P. (1981). *The physicists*. Macmillan, London.
Squires, Euan (1986). *The mystery of the quantum world*. Adam Hilger, Bristol.
Stewart, Ian (1989). *Does God play dice? — the mathematics of chaos*. Basil Blackwell, Oxford.
Zukav, Gary (1980). *The dancing Wu Li masters*. Bantam, New York.

RESEARCH PAPERS AND REVIEW ARTICLES

*Aspect, Alain (1976). Proposed experiment to test the nonseparability of quantum mechanics. *Physical Review D*, **14**, 1944.
Aspect, Alain, Grangier, Philippe, and Roger, Gérard (1981). Experimental tests of realistic local theories via Bell's theorem. *Physical Review Letters*, **47**, 460.
Aspect, Alain, Grangier, Philippe, and Roger, Gérard (1982). Experimental realization of Einstein–Podolsky–Rosen–Bohm *Gedankenexperiment*: a new violation of Bell's inequalities. *Physical Review Letters*, **49**, 91.
Aspect, Alain, Dalibard, Jean, and Roger, Gérard (1982). Experimental test of Bell's inequalities using time-varying analyzers. *Physical Review Letters*, **49**, 1804.
Baggott, Jim (1990). Quantum mechanics and the nature of physical reality. *Journal of Chemical Education*, **67**, 638.
*Bell, J.S. (1964). On the Einstein–Podolsky–Rosen paradox. *Physics*, **1**, 195.
*Bell, J.S. (1966). On the problem of hidden variables in quantum mechanics. *Reviews of Modern Physics*, **38**, 447.
Bell, John (1990). Against 'measurement'. *Physics World*, **3**, 33. See also the reply from Gottfried, Kurt (1991). *Physics World*, **4**, 34.
*Bohm, David (1952). A suggested interpretation of the quantum theory in terms of 'hidden' variables, I and II. *Physical Review*, **85**, 166.
*Bohr, N. (1935). Can quantum-mechanical description of physical reality be considered complete? *Physical Review*, **48**, 696.
Clauser, John F. and Shimony, Abner (1978). Bell's theorem: experimental tests and implications. *Reports on Progress in Physics*, **41**, 1881.
*Clauser, John F., Horne, Michael A., Shimony, Abner, and Holt, Richard A. (1969). Proposed experiment to test local hidden-variable theories. *Physical Review Letters*, **23**, 880.
DeWitt, Bryce S. (1970). Quantum mechanics and reality. *Physics Today*, **23**, 155.

*Einstein, A., Podolsky, B., and Rosen, N. (1935). Can quantum-mechanical description of physical reality be considered complete? *Physical Review*, **47**, 777.

d'Espagnat, Bernard (1979). The quantum theory and reality. *Scientific American*, **241**, 128.

*Everett III, Hugh (1957). 'Relative state' formulation of quantum mechanics. *Reviews of Modern Physics*, **29**, 454.

*Freedman, Stuart J. and Clauser, John F. (1972). Experimental test of local hidden-variable theories. *Physical Review Letters*, **28**, 938.

Ghiradi, G. C., Rimini, A., and Weber, T. (1986). Unified dynamics for microscopic and macroscopic systems. *Physical Review D*, **34**, 470.

*Heisenberg, Werner (1927). The physical content of quantum kinematics and mechanics. *Zeitschrift für Physik*, **43**, 172.

*Schrödinger, Erwin (1980). The present situation in quantum mechanics: a translation of Schrödinger's 'cat paradox' paper, (trans. John D. Trimmer). *Proceedings of the American Philosophical Society*, **124**, 323.

Shimony, Abner (1986). The reality of the quantum world. *Scientific American*, **258**, 36.

*Wigner, Eugene (1961). Remarks on the mind–body question. In *The scientist speculates: an anthology of partly-baked ideas*, (ed. I. J. Good). Heinemann, London.

*These papers are reproduced in Wheeler, J. A. and Zurek, W. H. (eds.) (1983). *Quantum theory and measurement*. Princeton University Press.

Name index

The symbol *f* indicates footnote

Subject index